How to do a Y-DNA study
including Next Generation Sequencing

Which can contain atDNA, STR Y-DNA and SNP Y-DNA

Next Generation Sequencing (NGS) Y-DNA testing
with Family Tree DNA's Big-Y test
analyzed by YFull

By James Lee Rader

How to do a Y-DNA study
including Next Generation Sequencing

Which can contain atDNA, STR Y-DNA and SNP Y-DNA

Next Generation Sequencing (NGS)
Y-DNA testing with Family Tree DNA's Big-Y test
analyzed by YFull

James L. Rader

June 2017
Antelope, California

By James Lee Rader

First Printing: 2017 --- Wednesday, June 28, 2017

ISBN-13: 978-1548313623

ISBN-10: 1548313629
Your book has been assigned a CreateSpace ISBN.
Published by author
my books are available at http://www.lulu.com/spotlight/jim154

DEDICATION

I dedicate this book to those men who supplied the necessary DNA. They are; Alton C Rader **Age: 93,** Earl Francis Rader **Age: 78,** Harry Rader **Age:**95. I also include myself for comparison James Lee Rader **Age: 75**
It is obvious that without their samples this book would not exist!

Alton Clyde Rader

When Alton Clyde Rader was born in 1923 Kosciusko Co, IN, his father, Russell, was 32 and his mother, Hazel, was 33. He had one son and one daughter with Doris Eileen Mearing. He had two brothers and six sisters.

   Father: Russell Leo Rader (1891-1969)
   Mother: Hazel Devona Tullis (1890-1985)

He had one son and one daughter with Doris Eileen Mearing.
Doris Eileen Mearing (1921 - 1978)
   Children:
      1. Randy Lee Rader (1957-   )
      2. Nancy Jane Rader (1961-
Patricia Sue Rader Age: 58 (Born Oct 1958)
Randy Lee Rader Age: 59 (Born, 1957)

1945 WWII Navy

1942 Passenger USS Pastores 21 Feb 1942 from Norfolk, Va. to Bermuda 311 75 37 S2c
1942 Passenger USS Pastores 8 Mar 1942 Bermuda, BMI to NOB, Norfolk, Va 311 75 37 S2c
1943 USS Ranger 30 Dec 1943 311 75 37 AM2c enlisted 17Jun41 Detroit, Mich
1945 WWII Navy Muster service number 311 75 37 AM1C(T) 8-9/45
1945 Kentucky State Teachers College 1 Sep 1945 311 75 37 AM1C(T)
1945 Kentucky State Teachers College 1 Dec 1945 311 75 37 AM1c(T)

In early 45 I was in the A+R shop in Corpus Christi Texas NAS as Texas is not one of my favorite places I took tests, mental and physical to get rated the student Naval Aviator Program SNAP.

We went to Murray State teachers college in Ky, (an accredited course). The top half was sent to St. Marys in Calif. More courses plus lots of Military and faster, again the top half was sent to Pensacola Florida for pre-fight and finally flight training. Along came VJ day and the powers that be decided no more need for SNAP's

I worked with my father as a mason and carpenter, worked in and became a manager at a Brass and Aluminum Foundry for 18 – 20 years,

I then went to work for the State of Indiana at their Employment Dept. I worked my way up to Office Manager for 30 years, retired,

I then went to Sullivan College in Louisville Kentucky and came out as a pastry chef (Suma Cum Laude) at the age of early 70s

He had one son and one daughter with Doris Eileen Mearing.
Doris Eileen Mearing (1921 - 1978)
    Children:
        1. Randy Lee Rader (1957-   )
        2. Nancy Jane Rader (1961-
Patricia Sue Rader Age: 58 (Born Oct 1958)
Randy Lee Rader Age: 59 (Born, 1957)

Earl Francis Rader

When Earl Francis Rader was born on 1939, in Kern, California, his father, Oscar, was 31 and his mother, Ellen, was 30. He married Joan C Anderson on July 15, 1958, in Carson City, Nevada. They have two children. He has one brother and three sisters.
Scott Kimberlin Rader

Birth: 14 Jan 1939 • Kern, California; Residence: 1 Apr 1940 • Kern, California, USA; Marriage: 15 Jul 1958 • Carson City, Nevada
Earl Francis Rader married Joan C Anderson in Carson City, Nevada, on, when he was 19 years old.
His son Scott Kimberlin was born on 1959.
His daughter Kibby was born on, 1962, in Santa Clara, California.
AGE 44
Death of Father
His father Oscar Earl passed away on 1983, in Arroyo Grande, California, at the age of 76.

Earl graduated from Coalinga Community College with a degree in Business Administration. He then started working at Lockheed in San Jose for $1.95 per hour.

He was born in Bakersfield, California and went to school there. His dad's family was from Arkansas, and their life in the 30's with the depression was very difficult. Earl's grandfather was a particularly forward thinking man who built a school and a church in their small town.

A great memory from childhood was growing up in the small town of Avenal, CA near Bakersfield, where he played in track and football. He calls it "a town of caring people."

Earl's interest in retirement was to travel with his wife Joan in their RV, and they crossed the USA three times. They also enjoyed cruises.

Earl has always had a Christian faith, and as a child witnessed the energy, his family put into building the church in Arkansas. The church was later moved to Branson, Missouri. He also affiliated with the Boy Scouts. He thinks having chores as a child was good.

The best advice he could give is, "Do it now!!" He likes jokes and likes to laugh. He also would like to advise others to stay away from drugs. He does believe in carrying arms.

Earl's dad influenced him greatly. He would ride on his dad's shoulders to see what he was doing. He would follow his dad around if he was feeding chickens or whatever.

What Earl would like to be remembered for is that he worked for Lockheed for 41 years, and spent years working on the Polaris-Poseidon. He loved his volunteer work at Pebble Beach where he spent 2 weeks each year.

Harry Wayne Rader 1921–

When Harry Wayne Rader was born on 1921, in Tennessee, his father, Charles, was 31 and his mother, Martha, was 29. He married Agnes Rena Johnson in June 1951. They had three children during their marriage. He had four brothers.

He worked in Andy's sash lumber / burr corn Roller Mill where one of his responsibilities was to shell corn. Andy had invented a system which allowed him to shell the corn very fast. All Harry had to do was feed the ears of corn, one after another, into the sheller, letting the equipment do the work. He was paid 1 penny for each bushel of corn he shelled. Due to Andy's ingenious set-up of the equipment, many days Harry was able to shell 100 bushels of corn. This netted him a dollar a day "which was big money back then."

He was just 16 years old when his grandfather passed away in 1937. Andy would not live long enough to see Harry graduate from high school in 1939 nor when he left Dulaney to serve his country 4 years later.

Harry graduated from Greeneville High where he proved he was a very talented athlete. Among his many accomplishments, he was Captain of the Greeneville Greene Devils basketball team during his senior year.

Harry served his country during WWII. He enlisted in the Army on 12 Jan 1943 at Fort Oglethorpe in Georgia. He was a Sergeant in the Battery B 469th Antiaircraft Artillery AW Battalion. He was Chief of Section 2601 where he served as a rifle marksman.

He received many decorations and citations for his service. These included the World War II Victory Ribbon, Good Conduct Medal, America Theater Ribbon, AP Theater Ribbon with 4 Bronze Service Stars and 1 Bronze Arrowhead, the Philippine Liberation Ribbon and many others.

Harry was discharged from the service on 27 Dec 1945 at Camp Chaffee in Arkansas. He served a total of 2 years, 11 months and 23 days of which 2 years, 3 months and 18 days were spent in Foreign Service.

He had 4 brothers, John 'Lyle,' Charles 'Willis,' Clyde 'Ray' and James Richard. James died as an infant. Lyle and Willis served in the military during the same time as Harry but returned on different dates. Ray, who was a senior in high school, died from injuries he received in a car accident while Harry was serving overseas and could not return for the funeral. This was a very difficult time for the Rader household, especially for the parents of the 5 sons.

Harry lost his father, Charlie, to cancer less than 2 years after he returned home from the service. Ironically, he lost his mother, Martha, 32 years to the day and basically to the minute, after his father passed away.

Harry returned to school, specifically technical school, on the GI Bill following his years in the service. He specialized in agriculture which assisted him in becoming a very successful farmer during the many decades that followed.

Harry has also excelled as a community leader since his return from the service. He is a charter and Life Long member of the Andrew Johnson Veterans of Foreign Wars Post 1990, a 'Free and Accepted Member of Masons', Lodge #463 and served as President of the Glenwood Ruritan Club.

Harry has been a hard worker all his life. He worked in the construction field as an iron worker for over 35 years. He was employed by

TVA and a variety of construction companies, performing the ironwork in numerous buildings and other structure throughout East TN, including the highway bridges along Interstate 81. He assisted in building one of the additions to his childhood's home church, Rader's Chapel in Dulaney and the addition to the Cedar Hill CP Church.

His long days at his construction job did not end when he arrived home from work for he continued his duties as a farmer. He raised a variety crops including tobacco and hay. The stock animals included cattle, swine, and poultry. His day usually began well before sun-up and ended after the sun went down.

Harry also married, raised a family, assisted in caring for his elderly mother and continued to remain very active in the church during this time. Even today, at the age of 93, he attends church and even continues to mow his yard every week. In May 2014, he was recognized for his service in WWII when he was presented the 'Quilt of Valor.'

Though Harry has accomplished much in his life, he is better known for his simple demonstration of his Christian faith. He begins each morning in prayer and studying the scriptures. He believes in the Golden Rule, and he is a prime example of the Serenity Prayer for he accepts the things he can not change, he has the courage to change the things he can, and he does possess the wisdom to know the difference. I am blessed and thankful to have him as my earthly father.

James Lee Rader

When James Lee Rader was born in 1942, in Oxnard, California, his father, Thomas, was 31 and his mother, Evelyn, was 31. He married Carol Raye Tucker, and they had two children together. He then married Priscilla Long in 2009, in Antelope, California. He had three brothers.
Parents Thomas Glenn Rader and Evelyn Lanore Stevenson

AGE 18 Surveyor Highways Aug 1960 – Aug 61• Baker, San Bernardino, CA

James Lee Rader married Carol Raye Tucker in Las Vegas, Nevada, on March 2, 1963, when he was 20 years old. Carol Raye Tucker 1944–2008

AGE 22 Death of Father His father Thomas Glenn passed away on June 24, 1965, in Seattle, Washington, at the age of 54. Thomas Glenn Rader 1910–1965

AGE 24 Surveyor DWR 03 Apr 1967 – Jul 73• Victorville, San Bernardino, CA

AGE 29 Death of Mother His mother Evelyn Lanore passed away on November 2, 1971, in Victorville, California, at the age of 60. Evelyn Lanore Stevenson 1911–1971

AGE 30 Inspector DWR Jul 1973 – Jul 76 • Perris Dam
AGE 34 Analyst DWR 21 Jun 1976 – Jun 83• Sacramento, Sacramento Co, CA

AGE 40 Death of Brother His brother Guy Steven died on 1982, in Santa Cruz, California, when James Lee was 40 years old. Guy Steven Rader 1938–1982

AGE 40 Programmer EDD 01 Jun 1983 – Feb 85• Sacramento, Sacramento Co, CA
AGE 50 Investor 22 Jul 1992 – Jan 00• Sacramento, Sacramento Co, CA

AGE 65 Death of Wife His wife Carol Raye passed away in 2008, in Sacramento, California, at the age of 63. They had been married 44 years. Carol Raye Tucker 1944–2008

AGE 66 Death of Daughter His daughter Debra Anne passed away in 2008, in Sacramento, California, at the age of 43. Debra Anne Rader 1965–2008

AGE 66 Marriage James Lee Rader married Priscilla Long in Antelope, California, on 2009, when he was 66 years old.

Contents

How to do a Y-DNA study

How to do a Y-DNA study

What did this process produce? Why would a genealogist spend this amount of time and money?

Before this Study, I was looking for information on my immigrant ancestors specific village of origin, but after 30 years I still do not know where he came from!

I also took my first Y-DNA test in 2002, and no one matched me. I then said what many of you say; "I will wait for someone to take the test who matches me!" It is now 15 years later, and the only matches I have are people who I found with genealogy. I had to ask them to take the test, and the only ones who would agree to the test did so after I volunteered to pay for the test and manage the test for them!

Now after this Study **I now know the exact Y-DNA of my immigrant ancestor on one of my ancestor lines. Much like a surname I know his Y-DNA terminal SNP name.** Anyone in the world with that SNP in his DNA is from the same closely related group. I do not need his village in Europe. I must find men of the world with that SNP.

I also know each SNP that occurred before his time of about 300 years before the present time. There is another one each 120 years back in time, much like an ancestor chart. When someone matches one of those many SNPs I know he is a close relative, and I know when our common ancestor lived.

As we continue to develop our Genealogy, we get more serious about accuracy. The Chart below shows my Great Great Grandparents. How many surnames are represented in that group?   There are eight and were you going to progress a few more generations I would first have 16, then 32, and so on.

I am lucky, as my female friends tell me; I have a Y chromosome which I can test! How many of my eight ancestor lines does my Y chromosome represent? Only my "Rader" Rader male ancestors contain that Y-DNA. If I want to study all eight I must recruit seven more men who carry each of the needed Surnames

I will bet that many of you have tested your DNA at AncestryDNA.com. Your third cousin or fourth cousin who shares 100-300 centimorgans with you would be good candidates for you to build a relationship. You may need to ask the female who you match about testing her brother. That is one way to find those males who carry the surname you want to research.

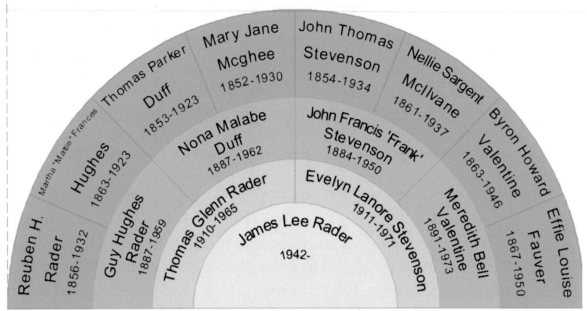

Another way is to do reverse genealogy! Look at your tree which you have on Ancestry. Do not look for ancestors, look for siblings. Then look for children! Repeat until you get to living people with the surname you want to prove. Several who each contain the same surname would make a fine study group. If you only use two and their DNA does not match you will need the third to break the tie.

Chapter 1 – 1980 who are we ?– 2012 One Name Study

My study of the Rader Surname or should I say my Rader ancestry began with a question from one niece. Where do we come from? And I naïvely said I should be able to figure that out. I have traced our ancestry in America but know very little of its origin. Our Rader surnamw immigrant had a German Bible in his estate inventory, and he lived in the German speaking regions of colonial America! It's now been 30 years, and I still don't know where Casper Rader came from!

Where did I actually start and what did I do to get my current Knowledge. I started asking relatives family members and others where my family came from. After talking to my father's sister multiple times, she said: "Oh you don't know that we came from Greenville, Tennessee." And furthermore, there is a living cousin back there who is studying the family.

With that, I contacted Margaret Rader from Greenville Tennessee. She liked to tell the story of her father calling her whenever a stranger came into town asking about Rader's. He would say "Sister go get your book." Her book was a three-ring binder about 3 inches thick containing family group sheets. After she provided me a copy of her information, I looked for a computer program to manage it.

In the early 1988 Roots III was the first database program I used, and it was followed by many others over the years. As I entered Margaret's data, it became obvious that the names of the people were not unique. There are many John's and Henry's and Williams and many born in the same era. It became very difficult to separate them and keep straight and which one was on which tree. The data wasn't complete, so no one knew for sure how many trees there were.

Over the years that followed I attempted to create a complete database of all people who used the name Rader in the world. When the 1880 US Census became available on CD, I extracted all of the Rader families in that Census and entered them in my database. I also harvested many databases from the rootsweb.com website and added them to my database. By using the duplicates tool on the software, I merged redundant entries for the same person. I still could not correctly identify the Rader families in the period before the Civil War, who lived in Greene County Tennessee.

To get more people involved, I published the material I had collected. That effort culminated with the publishing in 1995 of my multivolume work- "Second Attempt to collect ALL of the Roder, Reader, Raeder, Röder, Roeder, Ritter families in America." (It is available on kindle) I also loaded most of this genealogy on most of the online databases. Over the years since, I've added to the database and corrected many things that were duplicates and errors in the database. I also used tools like ancestry.com to add siblings and other family members. I included my other great great great grandparents and also included all of their descendants of that effort to find living cousins. The database today has over 100,000 people in it.

In the year 2002, I took my first DNA test. And as I attempted to understand the use of the DNA as a tool for genealogy I took many more tests. I have been taking DNA tests for 14 years now and over 25 tests with many companies.

At this point, I looked at land records and probate records to identify which John Rader belonged to which family. In 2003 I hired a local researcher from Greene County Tennessee to map all of the deeds. He plotted them on quad maps so we could figure out where they lived in Greene County. That also helped separate them and collect them properly.

One Name Study Published in Journal of One-Name Studies, 2012

An American's approach to a ONS, or would this be called a "Genealogy"?

By James L. Rader jim@rader.org *www.rader.org*
Published in *Journal of One-Name Studies, October—December 2012*
http://www.one-name.org/journal/pdfs/vol11-4_full_Rader.pdf

What Surname am I following?

The primary focus of this One Name Study (ONS) is the families with names which sound like Rader, Roder, Rotter, Roeder, Roetter, Rather. I find these spellings are quite interchangeable in the records of the early 18th century.

When I work with original handwritten records, I find that there is another ingredient in the SURNAME variant discussion. Interpreting the handwriting can be difficult because many examples of cursive handwriting show that the author took little effort in forming the vowels.

In my book "Casper Rader" 1732-1812 Wythe County, Virginia (ISBN-13: 978-0615182179) I collected every source record available. This collection is included in the first 30 pages of the book. Casper paid taxes each year for 20 years on the same piece of Pennsylvania land. His name is spelled; Gasper Reedar, Gaspar Reeder, Gasper Reeder, Casper Rheder, Gasper Rheder, Casper Reader, Gaspor Reador, Casper Reeder,

Casper Rhoder, Gasper Rader. Would you call these variants or sloppy writing?

His actual signature from the actual scanned in source documents will bring another slant to the discussion. The first is from the actual parchment he signed in his "Oath of Allegiance" to the King of England when he landed in Philadelphia in August of 1750 at an age of 18.

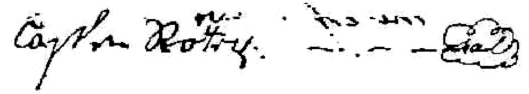

The second example is from his 1812 will, the actual will written in pencil stored in the probate court documents tied with a pink ribbon

Where do these Raders originate?

The simple answer is Europe, but that does not help us as we want to know a specific place.

Short History of the Rotters

Original German language by Gernot Rotter

English written by James L Rader

The names in this collection seem to be derived either from a profession, a location or an attribute. In the Germanic languages, the color Red is "Rot" so someone with the attribute of red hair would be a Rotter. There are also professions Roder (a wheel maker) and Roder (a clearer of the forest). So one would assume that as surnames were acquired, they were done simultaneously in many different locations. There is also a place called the Red Castle in an area of the Czech Republic, and one would assume the people of this castle were also called Rotters. The Rotter spelling seems most common in Austria today!

History of my project

I spent the last 20 years searching for

By James Lee Rader
Rader/Röder/Rötter families. That included several research trips to Pennsylvania, Virginia, and Tennessee. To produce my book "Casper Rader 1732-1812 Wythe County, Virginia" (ISBN-13: 978-0615182179) I needed to decipher the land ownership which required an expert on the court cases in Greene County, Tennessee. That small part of the project cost over $3,000. I have produced more than ten books on the parts of the exploits of the families.

This book is the first draft of a Translation of a work from 1908. It was originally published in Old German script in a collection of genealogies titled "Genealogisches Handbuch Bürgerlicher Familien" Volume 5 by Bernard Koerner which in English would be "Family Lineage Book."

This work is about the Roeder families from Schmalkalden in Thüringen. In 1317 this Rode (Rota, Rodo, Rotha) [cleared land] in mentioned as a lien of the Hennebergs. In the year 1333 Hermann Schultheiß. In 1348 Conrad of Rode (Kunz od Rode, of Rotha, Consze of Rota) becomes the Henneberg's "Steward at Frankenberg."

Roeder from Schmalkalden in Thüringen

James L Rader

This original work is available in the public domain, but it is in the printed old script and uses very archaic terms. I carried a copy of this work for over 20 years before finding a German translator willing to translate it. This translation work was begun by Michael Mayer-Kielman of Wilton, California, USA and completed by Kayla Rush krush728@gmail.com. The total cost of translation for the 62 pages came to $900.

I have also produced a new book which presents the project and includes European records and places identified which still need more research. The book is "Rotter / Rötter in early Europe" by James L Rader ISBN 5 800059 221839. This book will encourage others to get Y-DNA tests and do more European Genealogical research

Chapter 2 – You should use DNA

like any other type of record to prove your research!

When you're researching your ancestors, you are using genealogical tools. Those tools include many different record sets; land records, census records, burial records, and many other types of records. You can now add to that list DNA records or DNA tests.

The important point is that in our genealogical research

We should be using all available records.

What do you want to prove?

Each test has the capability of proving various things based on the relationships they show.

Some DNA shows close relationships. For example, a child will have one-half of the autosomal DNA of a parent. Every child has their mother's mtDNA. Every son will have his father's Y-DNA. A daughter will have 2 X chromosomes from her mother. A son will have one X chromosome from his mother.

Which type of DNA will help?

atDNA – parents give to all

mtDNA – mothers give to all

Y-DNA – Fathers give to sons

X-DNA – 1 or 2 from which

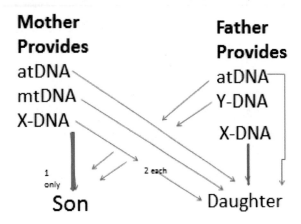

mtDNA, for example, is very good at disproving the relationships. Y-DNA can specifically prove a male lineage. An autosomal DNA can prove recent relationships.

That includes Y-STR, Y-SNP testing,

If you have been around DNA for a while, you know about the Y-DNA testing. The type you probably are familiar with is where we determine the number of STRs at each of 37 locations. With the new test, we will measure SNPs. I will give you much more detail about the new SNP measuring and the big Y-DNA test from family tree DNA in a later section.

Autosomal DNA testing (atDNA)

Ancestry's "everything test" is an autosomal test much like the autosomal test called relative finder at family tree DNA and the 23andMe test.

The fact that you get a half of your atDNA from each parent, who also got a half from each of their parents means you have one-quarter only from your grandparents. So when you look back several generations, you will see you quickly become diluted to where you have less than 1% and sometimes none. My fourth cousins, whom I will describe later, only give me a very small piece. And two out of three don't give me any at all.

How to do a Y-DNA study
Mitochondrial DNA (mtDNA) testing where appropriate

mtDNA testing has been around for a long time, and the anthropologists have been using it, they started using mtDNA. Its weakness is its size. It is very small and has very little DNA. If you come from a unique group, that can be very helpful. However, if you come from standard European stock, type R1b, you will match everybody. My complete mapping proves that I am related to 50% of the women in Europe.

As genetic genealogists, we must keep up with all the latest developments and make use of whatever tests are available. The technology available to us increases every year. At this moment you must buy STR tests separately from SNP tests. But soon you will get them together in one package at a much cheaper rate.

The quality of the tests is increasing also. The current methods cut the DNA into very small pieces, analyze them and put them back together. As we move forward and get more computer power, we can use much larger pieces which will decrease the errors at the cut points.

The big limitation, especially with Y-DNA testing, is actually to find appropriate candidates to test. You usually must pay for the test because the person you must test has zero interest.

Back to whether you should take a test? You do not have all of the DNA which you would like to sample. When you look at your great-great-great-grandparents level, you will see you have about 16 people. They represent 16 different surnames. If you would like to assure yourself that you have accurate research on those surnames, you must do Y-DNA testing on each surname on a male with that surname.

How accurate are these tests?

The original gold standard test to prove relationships was the Y-DNA 37 marker test. We concluded that test 10 years ago with three of us. There have been many tests since then, and most did not seem to help with the genealogy as I was trying to practice it. They were used to figure out living cousins.

Ancestry.com's test has turned out to be the most popular of that type test. They have more than 4 million people in their database, and it is a very good place to find living cousins. The obvious best candidate for that tool would be an adoptee, trying to find his family. It is working well for that purpose. It is also a good place to find those relatives who may have the type of DNA you need to complete your study.

The Y-DNA test has been increasing the available number of markers tested. In Casper's descendants case, the 37 marker tests are still quite useful. Our new pursuit has increased the number of markers to 111. But so far that has added no value. If you match on 37 markers exactly with those already tested, then you are related to each other very closely! You come from the same line of Y-DNA!

Autosomal DNA (atDNA) is looking for errors in your DNA. The term used for describing those errors is a snip (SNP). To find SNPs, you simply look at the backbone of the DNA and look for an error in the pattern. These errors are created by the father when he multiplies his DNA to put in his sperm. When he makes a mistake, and a child is created with that mistake in him, the child will pass it on to his children. These mistakes seem to happen about every 10th generation. The numbers are in dispute as to exactly how often they occur. It seems to vary from family to family.

So we look at these errors like a roadmap and also as a calendar. With that tool, we can figure out where and when an error occurred. We have been trying to develop a map of how the human race migrated with low-level DNA tests. With the newly improved test, we can measure in more detail and find more of these snips. Anthropologists are reworking their studies. It is expected that they will have new improved source data to analyze soon. Simultaneously the genealogist is tracing the more recent steps to map his genealogy.

Chapter 3 --What has Studying Y-DNA taught us

Before Y-DNA I felt that with enough genealogical research I could connect all Raders to one originating man. My analysis of the Rader Y-DNA project at FamilyTreeDNA.com first focuses on the Haplotype, which is the major category the test-taker exhibit. There are Types I, R, G, and E. This classification shows that the male originators of these four groups were not related for more than 10,000 years. These major groups further subdivide into subgroups which help me distinguish between closely related groups, but at this level, they are still not related in the time of surnames (over 1,000 years ago). I now have the 85 test-taker in 14 major groupings. These groupings are still not within the time of surnames. They are approximately within the past 3,000 years.

Surname -	Number tested	Haplogroup
Adam Roder 1645-	11	I-M223 I-L702
Casper Rotter 1732	7	R-M269 FGC3529
Ratz	1	E1b1b1 E-M35
	2	G-M201
I-CTS10228	7	I-M223 I-BY15562
I-M170	3	I-M253 I-M170 I-S2606
	7	I-M223 I-BY15562
	5	I-M253
J Haplogroup	1	J-M267
R1A	5	R-M512 R-M17 R-L1029
R1b1 not Casper	5	R-PH2647 R-L176 R-Y15784 R-U106
R1b1a2	21	R-M269 R-Z2103 R-S16361 R-Z198 R-DF21 R-BY11713 R-Z156
rater	7	I-M253
type E	5	E-M2 E-L117 E-M35
	85	

Gernot is haplogroup   R1b1a2a1a1b5
James is haplogroup   R1b1a2a1a1b5a

An examination of the tables below will show that the "a" on the end of the haplogroup translates into a genetic distance of 16 or 5,000 years

By James Lee Rader

Genetic Distance								Time to Most Recent Common Ancestor (Year

ID	modal	gernot	Mike	bob	Alton	Earl	James
modal	30	14	2	2	2	2	2
gernot	14	30	16	16	16	16	16
Mike	2	16	30	0	0	0	0
bob	2	16	0	30	0	0	0
Alton	2	16	0	0	30	0	0
Earl	2	16	0	0	0	30	0
James	2	16	0	0	0	0	30
Robert	3	16	1	1	1	1	1

Related	Probably Related	Possibly Related

ID	modal	gernot	Mike	bob	Alton	Earl	James
modal	30	4200	600	600	600	600	600
gernot	4200	30	5070	5070	5070	5070	5070	5
Mike	600	5070	30	150	150	150	150
bob	600	5070	150	30	150	150	150
Alton	600	5070	150	150	30	150	150
Earl	600	5070	150	150	150	30	150
James	600	5070	150	150	150	150	30
Robert	810	5070	360	360	360	360	360

0-270 Years	300-570 Years	600-870 Years	900-1170 Years

From
http://www.mymcgee.com/tools/yutility.html?mode=ftdnamode

Updating the calculation for 111 markers, we get 90 years instead of 150 years!

Time to Most Recent Common Ancestor (Years)								
ID	modal	Robert	Michael.	Alton		James	Harry		
modal	111	90	90	90		30	30	90	240
Robert Merl Rader Sr.	90	37	90	90	90	90	90	90	240
Michael David Rader	90	90	37	90	90	90	90	90	240
Alton Clyde Rader	90	90	90	37	90	90	90	90	240
Alton Rader	90	90	90	90	111	90	90	150	240
James Lee Rader	30	90	90	90	90	111	30	90	240
Harry Wayne Rader	30	90	90	90	90	30	111	90	240
Earl Francis Rader	90	90	90	90	150	90	90	111	240
Mr. Robert J Myers	240	240	240	240	240	240	240	240	37

0-270 Years	300-570 Years	600-870 Years	900-1170 Years

- Infinite allele mutation model is used
- Average mutation rate varies: 0.0026 to 0.0031
    Custom rates
- Values on the diagonal indicate number of markers tested

- Probability is 50% that the TMRCA is no longer than indicated
- Average generation: 30 years

The Casper Rader Y-DNA study

This is a DNA-Y **SNP** study which I did for the family of Casper Rader. Ten years ago we did a Y 37 **STR** DNA tests on three of the individuals in this test.

The original test concluded in 2006

The map below shows a portion of the eastern United States. The family which I'm trying to understand started with a person who got off the ship in 1750 in Philadelphia. He lived most of his life in Pennsylvania and in his retirement years he moved down the Shenandoah Valley to Wythe County. His children continued moving west in the search of land. Like most families, some children didn't stay with the family. His son Phillip married a Quaker girl who did not want to be around slavery, so they moved north. His son Daniel again looking for better land moved south to Alabama.

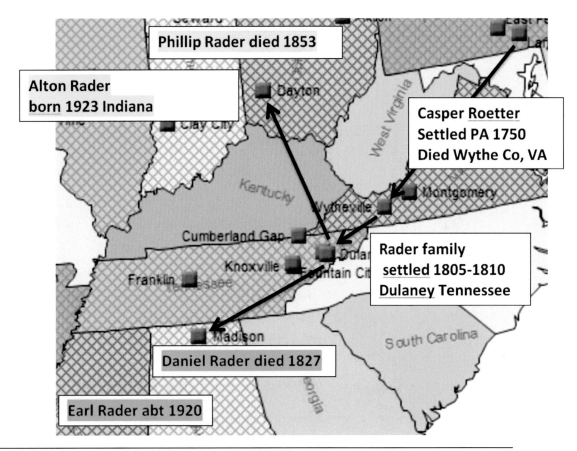

By James Lee Rader

Phillips line of Raders stayed in the North and is still in Indiana today. And we were lucky enough to recruit Alton who is a direct descendant of Philip.

Daniel's line continued to move west over the years, and we were lucky enough to find Earl whose family had been in the San Joaquin Valley of California for some time and is still in California.

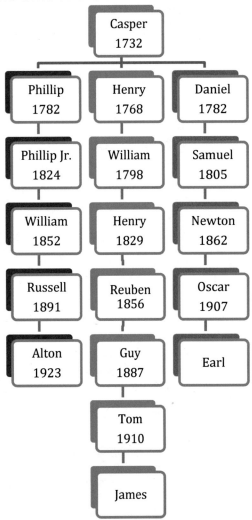

Our genealogy research was used to construct the chart.

The third person the study is me. My line was in the Greene County, Tennessee area with the family until just before the depression. My grandfather moved to Seattle Washington and then after the depression moved to Southern California. I live in Northern California.

So as you can see the three branches were not living next to each other for quite a bit of time. Is not likely that there was any exchange of DNA between the people living so far apart. So the plan was that if we can get Y-DNA from all three people, we could get a good idea of what the original immigrants Y-DNA was.

How to do a Y-DNA study

Ten years ago we collected samples so we could prove we were related to each other. Alton Rader from Indiana represents the descendants of Casper's son Phillip. Earl Rader from California represents the descendants Casper's son Daniel. And I, Jim Rader, represented the descendants of Casper's son Henry.

A 37 STR marker Y-DNA test was used to confirm our relatedness to each other. Science tells us that the Male Y-DNA passes from father to son unchanged. As you can tell from the chart, we selected male-to-male-to-male lines with no Rader females in the line. That was to test people who theoretically would have the same Y-DNA as Casper Rader from 1732. **The test results came back and indicated that we have identical Y-DNA at the 37 places where we were tested.**

Family Tree DNA help page states the following

Genetic Distance	Relationship	Interpretation
0	Very Tightly Related	A 37/37 match between two men who share a common surname (or variant) means they share a common male ancestor. Their relatedness is extremely close with the common ancestor predicted, 50% of the time, in five generations or less and over a 95% probability within eight generations. Very few people achieve this close level of a match. All confidence levels are well within the time frame that surnames were adopted in Western Europe.

https://www.familytreedna.com/learn/y-dna-testing/y-str/two-men-share-surname-genetic-distance-37-y-chromosome-str-markers-interpreted/

When an individual male takes a test today, I can tell them which of the major groups of Raders they belong. But further with the tools taught in Chris Pomery's class "Genetic Genealogy - DNA and Your Research (815)" http://www.dnaandfamilyhistory.com/courses I can tell them how closely related they are. In my ancestor's group, there is a group whose common ancestor is in the past 200 years or less and one other member who is over 5,000 years to a common ancestor. With that information, a new researcher can narrow his focus to a select group of Raders!

Chapter 4 you should not take a DNA test! Do a Y- STR study

What did this process produce? Why would a genealogist spend this amount of time and money?

Before this Study, I was looking for information on my immigrant ancestors specific village of origin, but after 30 years I still do not know where he was born!

I also took my first Y-DNA test in 2002, and no one matched me. I then said what many of you say; "I will wait for someone to take the test who matches me!" It is now 15 years later, and the only matches I have are people who I found with genealogy. I had to ask them to take the test, and the only ones who would agree to the test did so after I volunteered to pay for the test and manage the test for them!

Now after this Study **I now know the exact Y-DNA of my immigrant ancestor on one of my ancestor lines. Much like a surname I know his Y-DNA terminal SNP name.** Anyone in the world with that SNP in his DNA is from the same closely related group. I do not need his village in Europe. I must find men of the world with that SNP.

 I also know each SNP that occurred before his time of about 300 years before the present time. There is another one each 120 years back in time, much like an ancestor chart. When someone matches one of those many SNPs I know he is a close relative, and I know when our common ancestor lived.

As we continue to develop our Genealogy, we get more serious about accuracy. The Chart below shows my Great Great Grandparents. How many surnames are represented in that group? There are eight and were you going to progress a few more generations I would first have 16, then 32, and so on.

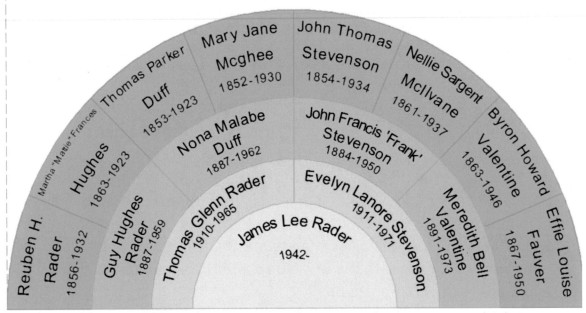

I am lucky, as my female friends tell me; I have a Y chromosome which I can test! How many of my eight ancestor lines does my Y chromosome represent? Only my "Rader" Rader male ancestors contain that Y-DNA. If I want to study all eight, I must recruit seven more men who carry each of the needed Surnames

I will bet that many of you have tested your DNA at AncestryDNA.com. Your third cousin or forth cousin who shares 100-300 centimorgans with you would be good candidates for you to build a relationship. You may need to ask the female who you match about testing her brother. That is one way to find those males who carry the surname you want to research.

Another way is to do reverse genealogy! Look at your tree which you have on Ancestry. Do not look for ancestors, look for siblings. Then look for children! Repeat until you get to living people with the surname you want to prove. Several who each contain the same surname would make a fine study group. If you only use two and their DNA does not match you will need the third to break the tie.

Chapter 5   FTDNA finishes the testing, and we have unique SNPs

FTDNA Match Date Search

James Lee Rader-Match Name	Shared Novel Variants	Known SNP Difference	Matching SNPs	Match Date
Earl Francis Rader	22	0	26,659	12/22/2016
Alton Rader	21	0	26,065	1/18/2017
Harry Wayne Rader	16	0	25,128	12/6/2016

1. A12555
2. FGC35929
3. FGC35931
4. FGC35933
5. FGC35936
6. FGC35937
7. FGC35938
8. FGC35939
9. FGC35940
10. FGC35941
11. FGC35942
12. FGC35943
13. FGC35944
14. FGC35945
15. FGC35946
16. FGC35947
17. FGC35951
18. FGC35952
19. ZS922

Chapter 6 --A progress report of the Y- STR study

What does testing more STR locations in Y-DNA give you?

Ten years ago we thought that a small number of STR locations would help you identify your relationship. We now know the 12 are 25 markers on the Y-DNA can only tell you if you are not related. If you don't match on all of those smaller numbers of markers, then it proves you are not related.

The 37 marker Y-DNA test is the gold standard. If you match on all 37 markers being tested, you are related within a short period. Then the question becomes should we enhance the number of markers being testing to get more accuracy. Moving out to the 67 and 111 marker level will show you the branching in your relationship the first son might have a difference in those levels while the second son did not.

Family Tree DNA Y- STR Results

The Y-DNA test we did initially.

Back in 2002, we had thoughts of what would work for figuring out your relatives using DNA. Brigham Young University came out with a plan where we would do 12 markers of the Y-DNA test and compare that to a four generation pedigree chart. We found out that was not adequate as we had already tested at 25 markers in November 2002. We found our 25 marker tests found many false positives. By early 2004 we had settled in on using 37 markers for a standard Y-DNA test. And in the 11 years since then, it is held up well for the Rader DNA study.

Rader DNA study

Adam Roder 1645	10 men type	I-M223
Casper Rotter 1732	6 men type	R-M269
I-CTS10228	10 men type	I
R1A	5 men type	R1A
R1b1 not casper	5 men type	R
R1b1a2 R-M269	13 men type	R-M
rater I-M253	5 men type	I-M253
type E	5 men type	E-L117

59 men tested - 8 subtypes

The big Y is also testing STR markers. At this time family Tree DNA is not using that data as they do not feel it is proven.

What am I testing in my Surname

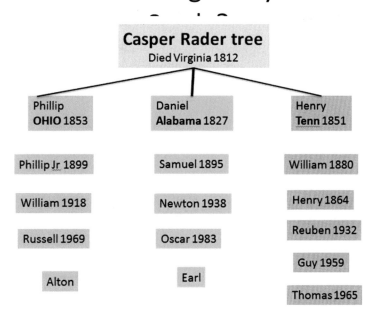

Casper Rader tree
Died Virginia 1812

Phillip OHIO 1853 — Daniel Alabama 1827 — Henry Tenn 1851

Phillip Jr 1899 — Samuel 1895 — William 1880

William 1918 — Newton 1938 — Henry 1864

Russell 1969 — Oscar 1983 — Reuben 1932

Alton — Earl — Guy 1959

Thomas 1965

Alton and Earl are Guy Rader's 5th cousins

The original FTDNA Rader Y-DNA Surname study

 Y-DNA - Standard Y-STR Values

PANEL 1 (1-12)					
Marker	DYS393	DYS390	DYS19**	DYS391	DYS385
Value	13	24	14	11	11-14

DYS426	DYS388	DYS439	DYS389I	DYS392	DYS389II***
12	12	12	13	13	29

PANEL 2 (13-25)					
Marker	DYS458	DYS459	DYS455	DYS454	DYS447
Value	17	9-9	11	11	25

By James Lee Rader

DYS437	DYS448	DYS449	DYS464
15	19	30	15-15-17-18

PANEL 3 (26-37)

Marker	DYS460	Y-GATA-H4	YCAII	DYS456	DYS607
Value	11	11	19-23	16	15

DYS576	DYS570	CDY	DYS442	DYS438
16	17	34-37	13	12

PANEL 4 (38-47)

Marker	DYS531	DYS578	DYF395S1	DYS590	DYS537
Value	11	9	16-16	8	10

DYS641	DYS472	DYF406S1	DYS511
10	8	10	11

PANEL 4 (48-60)

Marker	DYS425	DYS413	DYS557	DYS594	DYS436	DYS490
Value	12	23-23	16	10	12	10

DYS534	DYS450	DYS444	DYS481	DYS520	DYS446
16	8	11	22	20	14

PANEL 4 (61-67)

Marker	DYS617	DYS568	DYS487
Value	12	11	13

DYS572	DYS640	DYS492	DYS565
11	11	12	12

PANEL 5 (68-75)

Marker	DYS710	DYS485	DYS632	DYS495
Value	35	16	9	16

DYS540	DYS714	DYS716	DYS717
13	26	26	19

PANEL 5 (76-85)

Marker	DYS505	DYS556	DYS549	DYS589	DYS522
Value	12	11	12	12	11

DYS494	DYS533	DYS636	DYS575	DYS638
9	12	12	10	11

PANEL 5 (86-93)

Marker	DYS462	DYS452	DYS445	Y-GATA-A10
Value	11	30	12	12

DYS463	DYS441	Y-GGAAT-1B07	DYS525
24	13	12	10

PANEL 5 (94-102)

Marker	DYS712	DYS593	DYS650	DYS532
Value	22	15	20	13

DYS715	DYS504	DYS513	DYS561	DYS552
24	17	11	15	24

PANEL 5 (103-111)

Marker	DYS726	DYS635	DYS587	DYS643
Value	12	23	19	10

DYS497	DYS510	DYS434	DYS461	DYS435
14	17	9	12	11

MICRO ALLELES

Marker	Result	Normalized Result
DYS710	35.2	35

Rader Y surname study

Join The "Rader" Group Project
 https://www.familytreedna.com/group-join.aspx?Group=Rader
Surname Search Results

THE FOLLOWING NAMES MATCHED YOUR SEARCH REQUEST:	
NAME	**COUNT**
Rader	43

Surname Project Matches

If this Surname Project is associated with your last name, and we recommend that you join it! By joining your surname project, you can verify if you connect to one of the existing lines.

Click on the surname for project details or to join the project.

PROJECTS		
Project	**Mem bers**	**Description**
<u>Rader</u>	74	Rader, R360, Rotter, Rötter, Roeder, Raeder, Roder, Röder or...

PROJECTS		
Project	**Mem bers**	**Description**
<u>Germany-YDNA</u>	3316	**Welcome to the Germany-YDNA German Language Area DNA Re...**

If your surname is a variant of one of the above project(s), you can add your name to the project by clicking on the above link and ordering a test.

You can <u>order</u> a test outside the above-listed surname project(s).

Member Count 74
Project Website
https://www.familytreedna.com/groups/rader

Email
<u>Contact Group Administrator</u>

Description, this project includes all names which sound like Rader, Rotter, Rötter, Roeder, Raeder, Roder, Röder or other phonetic spelling variation of these surnames anywhere in the world, If you are a male named or are a female who knows such a male, please consider participating in our project. To find out more about our project, please contact:

Jim Rader at 916-366-6833 or jim@rader.org
www.rader.org/dna/indexM.htm

Requirements

A Surname Project traces members of a family that share a common surname. They are of the most interest in cultures where surnames are passed on from father to son like the Y-Chromosome. This project is for males taking a Y-Chromosome DNA (Y-DNA) test. The individual who tests must be a male who wants to check his direct paternal line (father's father's father's...) with a Y-DNA12, Y-DNA37, Y-DNA67, or Y-DNA111 test and who has one of the surnames listed for the project. Females do not carry their father's Y-DNA. Females who would like to check their father's direct paternal line can have a male relative with his surname order a Y-DNA test. Females can also order a mtDNA test for themselves such as the mtDNAPlus test or the mtFullSequence test and participate in a mtDNA project. Both men and women may take our autosomal Family Finder test to discover recent relationships across all family lines.

Surnames In This Project

Rader, Raeder, Reedar, Reeder, Reeter, Rider, Roder, Röder, Roeder, Rotter, Rötter, Ruder, Ryder

Y111-STR results differences table

Missing by one means that the probability of finding a common ancestor in 6 generations is 90%

	DYS589	DYS522	DYS494	DYS533	DYS636	DYS575	DYS638	DYS462	DYS452	DYS445	Y-GATA-A10	DYS463	DYS441	Y-GGAAT-1B07	DYS525	DYS712	DYS593	DYS650	DYS532	DYS715
Alton	12	11	9	12	12	10	11	11	30	12	12	24	13	12	10	21	15	20	13	24
Harry	12	11	9	12	12	10	11	11	30	12	12	24	13	12	10	22	15	20	13	24
James	12	11	9	12	12	10	11	11	30	12	12	24	13	12	10	22	15	20	13	24
Earl	12	11	9	12	13	10	11	11	30	12	12	24	13	12	10	22	15	20	13	24

Confidence

50%   90%   95%   99%

0   Very Tightly Related   A 111/111 match indicates a very close or im-mediate relationship. Most exact matches are 3rd cousins or closer, and over half are related within two generations (1st cousins).

2   4   5   6

1   Tightly Related   A 110/111 match indicates a close relationship. Most one-off matches are 5th or more recent cousins, and over half are 2nd cousins or closer.

3   6   7   9

2   Tightly Related   A 109/111 match indicates a close relationship. Most matches are 7th cousins or closer, and over half are 4th or more recent cousins.

5   8   9   11

3   Related   A 108/111 match indicates a genealogical relationship. Most matches at this level are related as 9th cousins or closer, and over half will be 5th or more recent cousins. This is well within the range of traditional genealogy.

6   10   11   14

Source https://www.familytreedna.com/learn/y-dna-testing/y-str/two-men-share-surname-genetic-distance-111-y-chromosome-str-markers-interpreted/

Y-DNA111 will Confidently, affirm a family group match between two or more men, and rule out (disprove) genealogical connections. Distinguish with the highest certainty possible the difference between recent and distant genealogical matches.

Y-DNA111: The Y-DNA111 test includes a balanced panel of sixty-seven Y-DNA STR markers, those from the Y-DNA67 plus forty-four more. The additional markers refine the predicted period in which two individuals are related. They eliminate unrelated matches. A close match at 111 markers indicates a common ancestor in recent generations, and an exact match indicates a close or immediate relationship.

Y-DNA111   results as displayed on individual page

Alton Rader

111 MARKERS – 3 MATCHES

GD	Name	DNA Haplogroup	Terminal SNP	Match Date
1	James Lee Rader	R-Y15646	Y15646	1/15/2017
1	Harry Wayne Rader	R-Y15784	Y15784	1/15/2017
2	Earl Francis Rader	R-Y15783	Y15783	1/15/2017

Harry Wayne Rader

GD	Name	Y-DNA Haplogroup	Terminal SNP	Match Date
0	James Lee Rader	R-Y15646	Y15646	12/11/2016
1	Alton Rader	R-M269		1/15/2017
1	Earl Francis Rader	R-Y15783	Y15783	12/21/2016

Earl Francis Rader

GD	Name	Y-DNA Haplogroup	Terminal SNP	Match Date
1	James Lee Rader	R-Y15646	Y15646	12/21/2016
1	Harry Wayne Rader	R-Y15784	Y15784	12/21/2016
2	Alton Rader	R-M269		1/15/2017

James Lee Rader

GD	Name	Y-DNA Haplogroup	Terminal SNP	Match Date
0	Harry Wayne	R-Y15784	Y15784	12/11/2016
1	Alton Rader	R-M269		1/15/2017
1	Earl Francis Rader	R-Y15783	Y15783	12/21/2016

GD = Genetic Distance

So what does this prove, we are Tightly Related. But Harry and Jim are Very Tightly Related.

By James Lee Rader

Out of 111 places tested we all match on 110. Earl has a different value at location DYS533, and Alton has a different match at location DYS712

STR matches and terminal SNPs

- STR tests sampled places on the Y chromosome where STRs were known to exist
- STR results → familiar matching lists of people with no more than 4/37, 7/67, 10/111 etc. mismatches
- SNPs commonly used to establish older kinship groupings – haplogroups & subclades
- Many testers wanted to know their terminal SNPs – the most recent, group-defining SNP they had (but mostly still much older than the period of interest to genealogists)
- Newer advances mean that long strings of the Y chromosome can be read and directly compared to other Y chromosomes. Now gives SNP testing far greater power to identify Y chromosomes that are closely related

Yfull.com is using the STRs from the Big-Y and is routinely reporting those 400 to 500 markers with their standard analysis.

January 16, 2017, Report on test results from the Descendants of Casper Rader Y-DNA study

Y-Full Report STRS 491

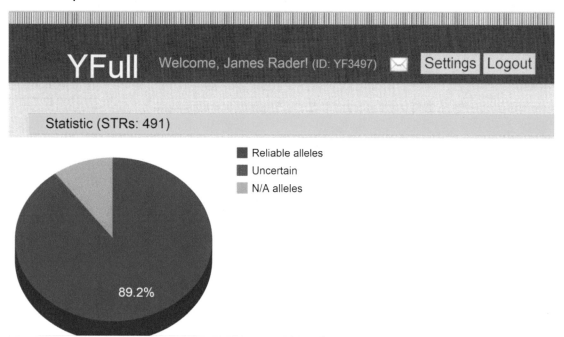

http://www.yfull.com/share/yreport/16fba0cce8053ba23623b0941c625e05

L

STR Matches

Compared STRs	Differences	Distance	ID	PM
	STR matches			
Close matches (0)		Distant matches (138)		
374	72	0.193	YF01632	✉
413	82	0.199	YF02155	✉
416	86	0.207	YF02328	✉
416	89	0.214	YF01954	✉
423	91	0.215	YF02444	✉

Chapter 7 YFull's report on STR matches from the Big-Y test

A new study suggests there is deterioration in the Y with age! Inspection of the table (below) which shows the results from my tests seems to support that. From my limited information, Harry is showing Age more than the other 3 of us!

THE HUMAN Y CHROMOSOME: AN EVOLUTIONARY MARKER COMES OF AGE Mark A. Jobling* and Chris Tyler-Smith published this month June 2017 Studies of genetic diseases show a strong bias towards fathers as the source of new mutations, and also show increasing mutation rate with paternal age (reviewed in REF. 12) 12. Crow, J. F. The origins, patterns, and implications of human spontaneous mutation. Nature Rev. Genet. 1, 40–47 (2000) https://web.stanford.edu/class/cs262/papers/humanYChromosomes.pdf

	Jim	Harry	Earl	Alton
Yfull #	3497	7939	8017	8525
age	75	96	77	93
STRS found	441	319	428	432
Length coverage	56.86%	49.15%	54.73%	52.89%
No base pair calls	44,786,696	46,765,436	45,332,628	45,806,604
BAM file size	0.69 Gb	0.55 Gb	0.67 Gb	0.78 Gb
Compared STRs	280-382	274-280	278-368	274-382

Select a sample
YF03497 (R-FGC35930) ▼ Submit

jim

STR matches

Close matches (3)   Distant matches (500)

Compared STRs	Differences	Distance	ID
280	3	0.011	YF07939
368	11	0.03	YF08017
382	14	0.037	YF08525

Harry

Earl
Alton

NOTE: STR matches based on "infinite alleles model"

Harry

Select a sample
YF07939 (R-FGC35930) ▼ Submit

STR matches

Close matches (3)   Distant matches (500)

Compared STRs	Differences	Distance	ID
280	3	0.011	YF03497
278	6	0.022	YF08017
274	7	0.026	YF08525

Jim

Earl
Alton

Earl	Select a sample

Select a sample
YF08017 (R-FGC35930) ▼ Submit

STR matches

Close matches (3)   Distant matches (500)

Compared STRs	Differences	Distance	ID
278	6	0.022	YF07939
355	9	0.025	YF08525
368	11	0.03	YF03497

Harry

Alton
Jim

Select a sample
YF08525 (R-FGC35930) ▼ Submit

STR matches

Close matches (3)   Distant matches (500)

Compared STRs	Differences	Distance	ID
355	9	0.025	YF08017
274	7	0.026	YF07939
382	14	0.037	YF03497

Alton

Earl
Harry

jim

By James Lee Rader

See Appendix for full list of STRs

Select a sample		
YF03497 (R-FGC35930) ▾	All ▾	

STR variants

Download .CSV

HG/SAMPLES	STRs	Mutation rate	ANC		DER
R-FGC35930	🔍 DYS501	★★★★★	8	→	7
R-FGC35930	🔍 DYS642	★★★★★	8	→	7
R-FGC35930	🔍 DYR97	★★★★★	12	→	13
R-FGC35930	🔍 DYR111	★★★★★	11	→	12
R-FGC35930	🔍 DYR77	★★★★★	12	→	11
R-FGC35930	🔍 DYS495	★★★★★	17	→	16
R-FGC35930	🔍 DYS511	★★★★★	10	→	11

Select a sample		
YF07939 (R-FGC35930) ▾	All ▾	

STR variants

Download .CSV

HG/SAMPLES	STRs	Mutation rate	ANC		DER
private mutation	🔍 DYS491	★★★★★	13	→	14
private mutation	🔍 DYS719	★★★★★	13	→	14
private mutation	🔍 DYR79	★★★★★	16	→	18
R-FGC35930	🔍 DYS501	★★★★★	8	→	7
R-FGC35930	🔍 DYS642	★★★★★	8	→	7
R-FGC35930	🔍 DYR97	★★★★★	12	→	13

Quote from FTDNA

The Big Y product is a Y-chromosome direct paternal lineage test. We have designed it to explore deep ancestral links on our common paternal tree. Big Y tests thousands of known branch markers and millions of places where there may be new branch markers. It is intended for expert users with an interest in advancing science.

It may also be of great interest to genealogy researchers of a specific line-age. It is not, however, a test for matching you to one or more men with the same surname in the way of our Y-DNA37 and other tests.

My experience shows that if you bring your own men to match, it becomes good for matching.

Chapter 8 –What do other analysts have to say?

A snip is an error in your DNA when your mother creates an egg, or your father creates a sperm, once in a great while they create an error. That error is designated by what's called a snip. It has a position and a value in the DNA. Over the years they've tried to get better at understanding SNPs and as you'll see they are considerably better.

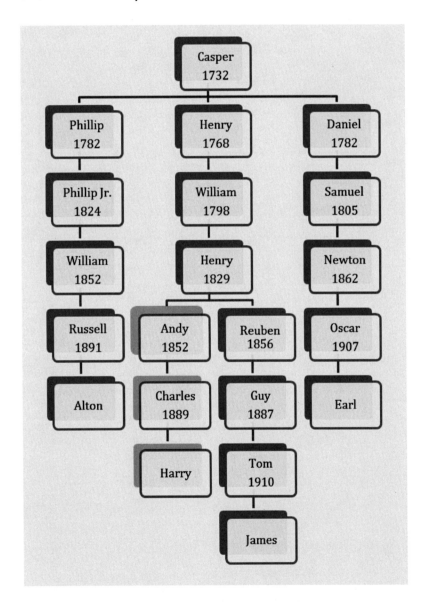

The science has moved on, and we have more tests available to us in 2016. We plan to use what is called NGS testing

The first tool for this study is extensive genealogy that shows your relationship to Casper Rader.

Secondly, someone needs to pay for the tests. The money to perform the needed tests and analysis. The cost for each person tested was about $900.00
The sale prices were

Test	Retail Price	Project Price	Sale Price
YDNA111	**$359.00**	**$339.00**	**$319.00**
Big Y	**$575.00**	**$575.00**	**$525.00**
Family Finder	**$79.00**	**$79.00**	**$59.00**
Analysis			**$50.00**

We welcome anyone who would like to participate. All you need to join is someone to test and the money to do the tests. If you would like to discuss your concerns with me just drop me an email at jim@rader.org.

And now the latest great breakthrough that we have is a tool to find all of the errors in your Y-DNA. This test is from family Tree DNA www.familytreedna.com, and it's called **the Big-Y**. I took that test in the spring of 2015, and when the results came back, they said well this is an experimental test, and we recommend other companies to help us interpret what it means.

There are three of them. I sent my BAM file which contains my Big-Y results to them and see what they come up with. The most productive group of those three is a company called Full www.yfull.com all which is out of Russia, and they have been quite productive

Y-Tree is producing a tree of matches

http://www.ytree.net/DisplayTree.php?blockID=30

Ytree with three results in and Red (appears darker in this B/W print) is used for men whose data has not yet been fully analyzed

Porter
N28007
Oakes
H2290
Williamson
196041

A darker (red) background is used for men whose data has not yet been fully analyzed. His position on the tree is not yet final, and will, in general, be downstream of the current position. He may not be positive for all the SNPs/INDELs in the block he descends from.
Men whose NGS data have been fully analyzed are indicated with a gray background color.

By James Lee Rader

Those men with a gray background and a pink bar to their left have been finalized, but haven't gone through as much scrutiny as earlier kits have. In particular, no search has been made for recurrent SNPs or other unusual mutations. For the vast majority of kits, this has no effect at all, but I will review them as necessary in the future.

A242
FGC18232?

A red (appears darker in this B/W print) SNP name with a '?' indicates that the SNP's status for the block is uncertain, but assumed to be negative. The same SNP probably occurs in a parallel block. It will be necessary to check BAM files or perhaps Sanger sequence some men to prove the result.

CTS4296
22233413-G-A

Mutations written with a red (appears darker in this B/W print) background fall within a region of the Y chromosome, such as the palindromic region, which has left the position of the mutation ambiguous. The true mutation may be at the indicated position, or at any one of a number of alternate positions.

With three test results in hand, they produced the following!

Z208 S362
CTS8289

S251 Z207
Z693

Z29720
CTS43
Z29721
Z29722
Z29723
Z29724
Z29725
Z29726
Z29727
Z29728
Z29729
Z29730
9845075-C-A
Z29731
Z29732
14031734-CTT-C
Z29733
Z29734
Z29735
CTS3444
CTS5447
Z29736
CTS5606
17201915-C-T
CTS7223
CTS7480
Z29737
Z29739
CTS10355
CTS10409
Z29740
Z29741
CTS10925
CTS11118
Z29743

Y15784 FGC35927
Y15646 FGC35935
Y15783 FGC35928

CTS10783
15607515-G-A
FGC41847
17342335-A-G
17577191-A-G
22444488-G-T
28801102-T-C

FGC35932
22468207-C-T
FGC35929
7908093-CAG-C
FGC35930
9027611-T-A
9027637-C-T
FGC35931
ZS9022
FGC35933
FGC35934
FGC35936
FGC35937
FGC35938
FGC35939
FGC35941
FGC35942
FGC35943
FGC35944
FGC35945
FGC35946
FGC35947
FGC35948
FGC35949
FGC35950
FGC35952

...ermez 8

Marimon
N113329
Vernade
N42387 DKAUK
Neal
PPT8C

1k Genomes
NA20342

1k Genomes
HG01148
1k Genomes
HG01464

Adams
133307
Palmer
366608

1k Genomes
NA19777

Rader
5185
Rader
557777
Rader
6589

Unknown
B26940

And adding the fourth test we got the following!

 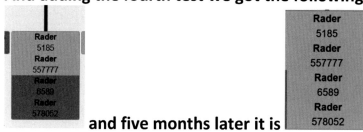 **and five months later it is**

30jan17 ytree

The instructions for help is from http://www.ytree.net/Instructions.html

Big Tree: Instructions

The Big Tree Project is open to any R-P312 man who wishes to contribute his NGS results. It's free. This project is one of many that relies on the open sharing of BigY or FGC results. These tests are useful only when the data is compared to those of other men.

There are two ways to send me your data:

1. E-mail your files directly (scotsgenealogy (at) gmail.com). Details of what to e-mail can be found below. Please don't forget to let me know your kit number.
2. I routinely scan the BigY and FGC upload folders of the R-P312, R-L21, R-DF27 and R-U152 Yahoo Groups. This is the easiest way for me to get your data. It's free to join these groups and upload your data. For R-L21 men, you can also e-mail your BigY results to M.W. Walsh (mwwdna (at) gmail.com), administrator of the R-L21 FTDNA Project, who can take care of uploading your results for you. If you upload your data to one project, please don't e-mail it to me.

Big Y

For the BigY, the files I need are the "Raw Data" files. This consists of a BED and VCF file which can be downloaded from the BigY results page. They come as one zip file, which is about 300kB or so in size. Please don't extract the files from the ZIP archive, leave them as they come.

Should you decide to upload them to a Yahoo project or e-mail them to me, the common format for labeling the file is:

Haplogroup_KitNumber_AncestralSurname_BigY_RawData_YYYYMMDD.zip
Example: DF21_196041_Williamson_BigY_RawData_20150325.zip

The surname that appears in the filename is the surname I'll use with your kit number on the tree. You can always e-mail me to make changes.

Your "Raw Data" is downloaded from your FTDNA page. After you sign in click on the link for your "Big Y Results."

Now click the blue "Download Raw Data" button in the upper right corner of the window.

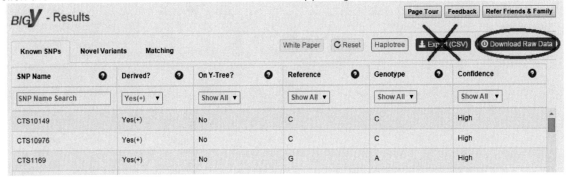

Finally, click the green "Download VCF" button in the windows the opens.

This will prompt you to save the .zip file to your computer. Please don't modify this file before uploading it or e-mailing it.

Big Y BAM File

The BAM file for BigY results is accessed in much the same way as the raw, VCF data. By first contacting FTDNA via https://www.familytreedna.com/contact.aspx#contactForm you can request access to the BAM file for your results. This file is the true raw data for the BigY test. It is a large file, generally about 1GB in size, and you need special software to read it. I rarely need your BAM file, but it is useful to refer to for some situations. If you have it available, please forward the link. If you are submitting BigY data to FullGenomes or YFull for further analysis, they will also need access to your BAM file.

http://www.ytree.net/DisplayTree.php?blockID=30

A progress report on using big Y results.

At this point all four of my test-takers (Harry, Alton, Earl and myself) have Big-Y results run through family tree DNA. I have also loaded the BAM files to Yfull.

In the spring of 2015, I took the big Y test myself. It was recommended you hire external consultants to analyze the results. I selected all three of the consultants and had them do their analysis. After I had completed that, I published the book "Y-DNA Jim Rader's multiple tests to understand his Y-DNA" (Dec 2015). http://www.lulu.com/shop/james-l-rader/jim-raders-y-dna/paperback/product-22611560.html At that point my understanding of what I had completed was very limited.

In the summer of 2016, I saw the Irish DNA work on YouTube, with snips and STRs. They had added the SNP analysis to bring archaeological timeframe ancient history into the current genealogical timeframe. To do that they showed that you needed to have a Y111 test and a big Y tests.

> DNA Lectures - Who Do You Think You Are
> Using SNP Testing & STRs to … (John Cleary) - Part 1
> https://www.youtube.com/watch?v=hlxvdayxZiI
> Using SNP Testing & STRs to … (John Cleary) - Part 2
> https://www.youtube.com/watch?v=6l5QARvhb38
> Using SNP Testing & STRs to … (John Cleary) - Part 3
> https://www.youtube.com/watch?v=pxexkvfus6w

So in the fall of 2016, I tried to get the samples from the "Rader"men who had been active in my 2006 Y37 test updated. With more effort, I contacted the three individuals who had been in my 2006 tests and one who is a newly discovered relative.

My Genealogy research shows me I am related to the three men who I am testing as follows:

1. Earl is a 4th cousin twice removed on my Rader line
2. Alton is a 4th cousin twice removed on my Rader line
3. Harry is a 2nd cousin once removed on my Rader line

> And a 4th cousin twice removed on my Bauer line
> And a 4th cousin twice removed on my Andes line

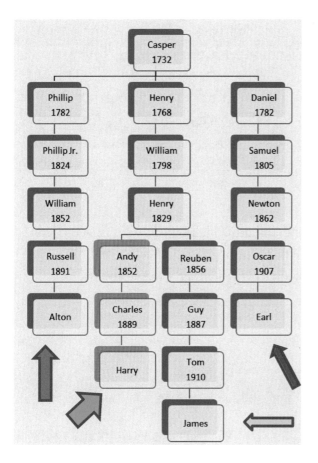

The current study which is now done

The science has moved on, and we have more tests available to us in 2016. We plan to use NGS testing

Big Y August 28, 2014

The Big Y product is a Y-chromosome direct paternal lineage test. We have designed it to explore deep ancestral links on our common paternal tree. Big Y tests thousands of known branch markers and millions of places where there may be new branch markers. It is intended for expert users with an interest in advancing science.

It may also be of great interest to genealogy researchers of a specific lineage. It is not, however, a test for matching you to one or more men with the same surname in the way of our Y-DNA37 and other tests.

www.familytreedna.com/learn/y-dna-testing/big-y/

By James Lee Rader

The announcement above was the hope of the test when announced two years ago. The Genealogy community has discovered more uses for this test.

The test is an SNP type test. Our DNA is subject to errors. When the male creates sperm, his copying process will "drop a stitch" make an error. That error will be passed on the every male descendant he creates. Those errors become like a milepost by which we can date stamp our ancestors. It happens on average every 150-200 years.

This test includes an additional person. I would like to include more if we can fund the cost and find the people who will provide their DNA for testing. We have Jim Rader's, Earl Rader's, Harry Rader and Alton Rader's completed.

NGS testing

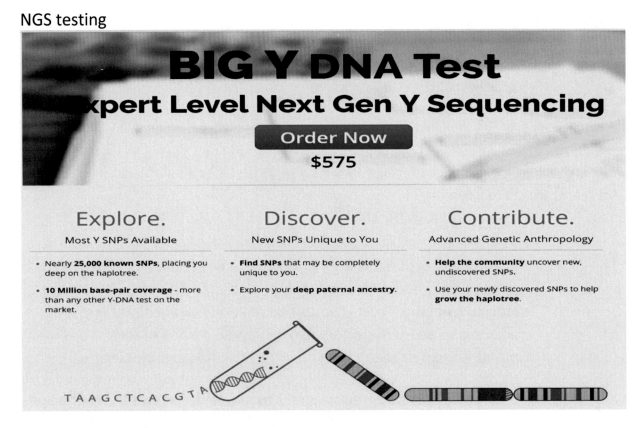

Big-Y results, do I understand this correctly?

These are the people involved in the 2016 Big-Y study

What can NGS testing prove?

I have a group of four men all who descended from the same person born in

1732. I have the big Y results for all four men. These four men match on Y-111 STR testing at least 110 out of 111 STRs.

The genealogy research on these four men is quite thorough. It has been researched by multiple researchers who publish results over the last 50 years.

It Has been confirmed by extensive land of probate research.
The table below entitled "compare novel variants all four 28Jan17" displays a lot of details obtained from that study.

The left column which have FGC names Yseq are the locations which I had tested with Sanger test methods to verify that the results from the big Y were accurate.

The second column is the name these particular locations are known by, at some companies.

The next section is the results of my big Y tests. The numbers are the physical location being tested.

The next section is Harry's test results followed by the blue highlighted section with Earl's test results and the final section on the right-hand side is Alton's results

I aligned them, so the places where there are blank spaces are locations where we do not match! You will see we do not match an equal amount or in the same places. Jim has 23 unique variants; Harry has 16 variants, Earl has 28, and Alton 22.

compare novel variants all 4 yfull 18apr2017

SNP name	Jim	Harry	Earl	Alton
			13599179	
FGC35933	14148053		14148053	
FGC35934	14330402		14330402	14330402
FGC35936	14488481			14488481
FGC35937	14766231	14766231	14766231	14766231
FGC35938	15449620	15449620	15449620	15449620
FGC35941	15942532	15942532	15942532	15942532
FGC35942	16856500	16856500	16856500	16856500
FGC35943	17417247	17417247	17417247	17417247
FGC35944	17517692	17517692	17517692	17517692
			17887770	
FGC35945	18032596	18032596	18032596	18032596
FGC35946	18122933	18122933	18122933	18122933
			18396980	
FGC35947	18693670	18693670	18693670	
Z688	22239121		22239121	22239121
FGC35948	22271952	22271952	22271952	22271952
FGC35949	22465458	22465458	22465458	22465458
FGC35949	23353191	23353191	23353191	23353191
FGC35952	23392463	23392463	23392463	23392463
FGC35929	7905834	7905834	7905834	7905834
FGC35930	8374882		8374882	8374882
YFS450776	9027611		9027611	9027611
YFS450777	9027637		9027637	9027637
FGC35931	9123622	9123622	9123622	9123622
ZS9022	9378241	9378241	9378241	9378241

I assume that any snip they all have would be a snip the original man had. I don't know when three out of four have a snip how could that happen? Harry's test is lacking many matches that the other three have. If the original man Casper had them and Jim who is downline from Harry has them, then Harry must also have them!

One guess is there is a mistake in testing and test missed that particular snip. Another guess would be that when his father created his sperm, he made an error at that same location which caused the SNP to flip back to the original value. The problem with that is that seven positions do not flip in 120 years (the time between Henry 1829 and Jim 1942)

Compare novel variants all 4

As a test is completed, I also submit the tests for analysis to the same three companies we used before. The Russian company Yfull seems to be the best at putting trees together.

The most common question I hear from people my lectures is "what test should I take." My answers have been, for some time, what would you like to know? The second part of that answer is it's probably not you that must test.

Your DNA is a powerful tool that can help you prove or disprove many things about you and your relationships. It can also do the same thing between your cousins and your ancestors.

Why do you want to know the rest of your SNP errors?

The SNP technology is a whole different tool. When you're looking at a Y-DNA test, each SNP occurs in a certain order. They occurred in different births. And it's estimated they happen about once every 200 years. So the more of them you detect the closer you get to today's dates. My current big Y tests are showing a snip which occurred 400 years ago. So that is the time of surnames or as we call it genealogy.

Will you know where these people lived when the SNPs happened?

Who has that DNA?

The method used to determine where a SNP occurred is in the realm of archaeology. As they discover human remains, they use carbon dating and other tools to determine when they died. And they use a location where the remains were found to determine where.

So we are all waiting on archaeologists to dig up our ancestor's graves and determine which snips they had to give us that information

■ ■

Chapter 9 - - Big-Y Introduction

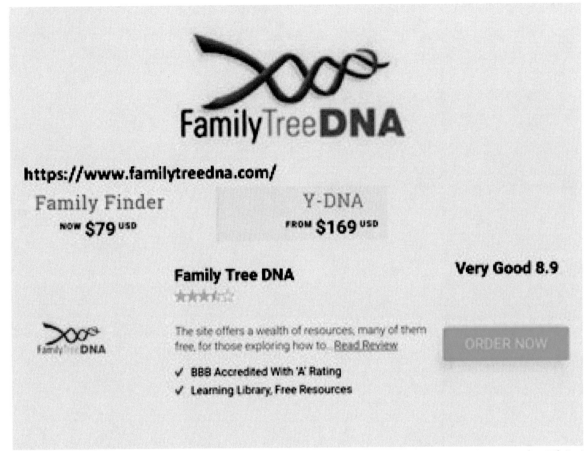

It is a Next-Generation study that uses the next generation testing tools. This is the second test of this group of individuals.

The Goal

Use NGS tests to define the tree of your group

Which will help you discover their history

Goal of Big Y test

Read virtually all of the Y chromosome that is useful for Genealogy
Try to reach back in time for connections
Connect outside the limits of Y-DNA STR tests
Supply the points between
    500 BC

1500s AD

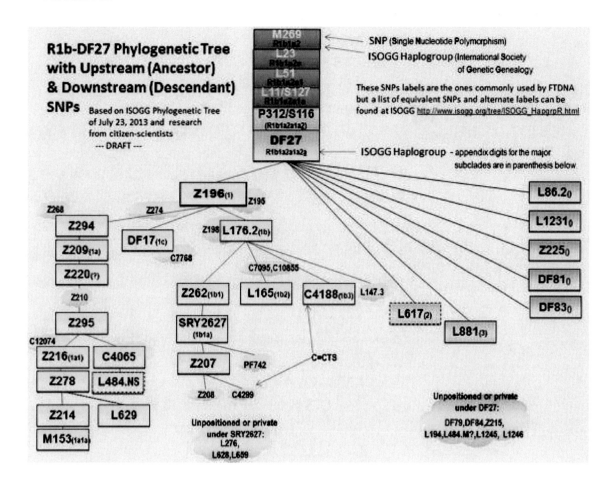

Detailed look at SNPs

R-Z206 Z206/S235 * Z690 formed 3500 ybp, TMRCA 3500 ybp
  R-Z206*
       id:YF01632
       id:NA12716CEU
  R-CTS4549 CTS4549 formed 3500 ybp, TMRCA 2800 ybp
       id:YF04245ITA [IT-PA]
       id:HG01359CLM
  R-Z208 Z208/S362 * CTS8289 formed 3500 ybp, TMRCA 3500 ybp
    R-Z208*
       id:YF04553
    R-Z207 Z207/S251 * Z693 formed 3500 ybp, TMRCA 3500 ybp
      R-Z207*

R-CTS43 CTS43 * CTS5447 * CTS10925+7 SNPs
id:HG01148CLM
id:HG01464CLM
R-Y15783 Y15646 * Y15783 * FGC35927/Y15784
formed 3500 ybp, TMRCA 3400 ybp

BRANCH ID R-Y15783 (age: 3399 ybp)   Formula: (3399)/1
SAMPLE ID NUMBER OF SNPS COVERAGE (BP) FORMULA TO CORRECT SNPS NUMBER

CORRECTED NUMBER OF SNPS
FORMULA TO ESTIMATE AGE
AGE BY THIS LINE ONLY

YF03497   22.0   8056427   22.0/8056427*8467165 23.12
23.12*144.41+60   3399

id:YF03497
id:NA19777MXL
R-CTS9762 CTS9762 formed 3500 ybp, TMRCA 3100 ybp
R-CTS9762*
id:NA20342ASW
R-CTS4299 CTS4299 formed 3100 ybp,
TMRCA 3000 ybp
R-CTS4299*
id:YF02301
id:YF01954
id:HG00126GBR
R-FGC11245 FGC11245 formed 3000 ybp,
TMRCA 2700 ybp
id:YF03914CHE [CH-BE]
id:YF02155ENG

How does YFull determine "formed" age and "TMRCA," **and the related confidence intervals, of the subclades in its YTree?**

A: The following definitions and methodologies relate to the subclades in the YTree:

Subclade name: Each subclade name is highlighted in green.

SNPs "defining" a subclade: These are listed to the right of the subclade name (by SNP name, with additional SNP names in the grey-shaded pop-up: "X (a number) SNPs"). The SNP list for a subclade may change in the future as more samples are added to the YFull database, and new branches are added.

Subclade "formed" age: The TMRCA (time to the most recent common ancestor) of a subclade is used as the "formed" age of each branch of the subclade. Stated otherwise, the formed age of a branch is the same as the TMRCA of the "parent" subclade of that branch.

Determination of TMRCA for a subclade: The general rule is that the TMRCA of a subclade is equal to the average age (after rounding) shown in the yellow bar of the YTree "info" pop-up table for the subclade. In the situations where the general rule is not followed YFull will add an explanatory note at the bottom of the table. For an example, see the table for the I1-Z63 subclade.

Rounding rules: An age of less than 500 ybp is rounded to the nearest "25" (e.g., 381 becomes 375); an age of 500 to 1999 is rounded to the nearest "50" (e.g., 1477 becomes 1500); and an age of 2000 or more is rounded to the nearest "100" (e.g., 3160 becomes 3200).

Formed CI xx% yyyy <-> zzzz ybp, TMRCA CI aa% bbbb <-> cccc ybp: CI means "Confidence Interval." A confidence interval is an indicator of the precision of the YFull "formed" age and "TMRCA" data in the YTree. YFull developed its own statistical analysis computer script in order to calculate its confidence intervals.

Yellow Bar in "info" pop-up table: The "ybp" (years before the present) for the subclade is the average of the ages of the branches and samples (if any) highlighted in green in the Branch ID column, as shown in the yellow bar "Formula."

Number of SNPs column in "info" pop-up table: For a branch, the number in this column is the average of the numbers reported for the samples in the branch. For a sample, the number in this column is the total of the Known SNPs and Novel SNPs located between the subclade and the present. These SNPs are identified in the Age Estimation table.

Other columns in "info" pop-up table: Branch numbers are averages of the numbers given for the samples in the branch. The two formulas used in the table are discussed in the FAQ: What is YFull's age estimation methodology?

What is YFull's age estimation methodology?

A: YFull uses a methodology based on the research and analysis discussed in Defining a New Rate Constant for Y-Chromosome SNPs based on Full Sequencing Data by Adamov, Guryanov, Korzhavin, Tagankin, Urasin (2015).

The methodology is reflected in the Age Estimation table for the each analyzed sample and in the subclade age pop-up tables linked to the YTree.

The first step is to select and count reliable derived Known SNPs for a sample. The number of counted SNPs appears in both tables.

The following five criteria are used to select reliable SNPS:

1. The coordinates of the SNPs must fall within the combBED regions designed to select X-degenerate segments. The combBED area borders were formed by mutual overlapping BED files taken from the work of Poznik et al. (2013) (total length of 10.45 Mbp) and by the generalized BigY BED file (11.38 Mbp long), published in the BigY White Paper (2014). The result was 857 continuous segments of the Y-chromosome with a total length of 8,473,821 base pairs.

2. Insertions and deletions (called "Indels") are excluded, as are multiple nucleotide polymorphisms (SNPs with more than one base position).

3. Variants detected in more than five different "localizations" are excluded. "Localization" means a group of samples from the YFull database belonging to the same subclade and having derived allele nomination. In some cases, the same derived variants may be found in different subclades or different haplogroups because of mapping errors or because the standard reference sequence is based mainly on haplogroup R1b data and to a lesser extent on haplogroup G data. This causes some variants in some haplogroups to be ancestral instead of

derived. Although YFull established the "five different localizations" criterion empirically, the criterion is soft but believed to be effective.

4. SNPs with only one or two "reads" are excluded.

5. SNPs are excluded if the "read quality" is less than 90%. Quality is determined pursuant to YFull's proprietary SNP rating system. See the FAQ How does YFull determine the quality rating for my SNPs?

The Age Estimation table for each sample provides a high level of detail about the application of the selection criteria. Reliable Known and Novel SNPs are listed in the "+Known SNPS, " and "+Novels" columns of the table and SNPs not selected are listed in the "x Known SNPs" and "x Novels" columns, with details related to the five criteria.

The second step of the sample age determination methodology is explained in the YTree "info" pop-up tables for the YTree subclades. For each sample in a table, two formulas are applied to the number of SNPs for the sample. The first formula corrects the SNP count to an assumed (or corrected) count from the combBed bp coverage area and the second formula establishes the age of a sample based on the corrected count. The second formula uses an assumed mutation rate of 144.41 years (0.8178*10-9, which is the average of the mutation rates of the ancient Anzick-1 sample and of a group of known genealogies, and an assumed age of 60 years for living providers of YFull samples.

See also: How does YFull determine "formed" age and "TMRCA," and the related confidence intervals, of the subclades in its YTree?

Last updated on Jan. 14, 2016.

2016 update

YFull YFull My news Welcome, James Rader! (ID: YF3497)

12/09/2015 New sample YF04553 in subclade R-Z208*
12/09/2015 Subclade R-Z208 divided to R-Z207, R-Z208
10/15/2015 New sample YF04245 in subclade R-CTS4549
10/15/2015 Subclade R-CTS4549 added to R-Z206 with SNP CTS4549

08/19/2015 New sample YF03914 in subclade R-FGC11245
08/19/2015 Subclade R-FGC11245 added to R-CTS4299 with SNP FGC11245
06/06/2015 Subclade R-Y15783 added to R-Z208
                        with SNPs Y15646, Y15783, Y15784
06/06/2015 Terminal haplogroup of sample
                        YF03497 changed from R-Z208* to R-Y15783

clarifYDNA

-----Original Message-----
From: Chris Morley
Sent: Wednesday, May 13, 2015, 10:25 PM
To: jim@rader.org
Subject: Re: [General feedback] which result do you analyze and how fast
(Sent by jimrader, jim@rader.org)

Thanks, Jim.

Your analysis is complete and can be accessed from
https://www.clarifYDNA.com/reports (login required).

There are some symbols in the report which have not yet been added to
the table on page 2. I suggest you ignore the undocumented symbols for
now.

Your "leaf" on the BigY-implied tree is line 20 of page 3.

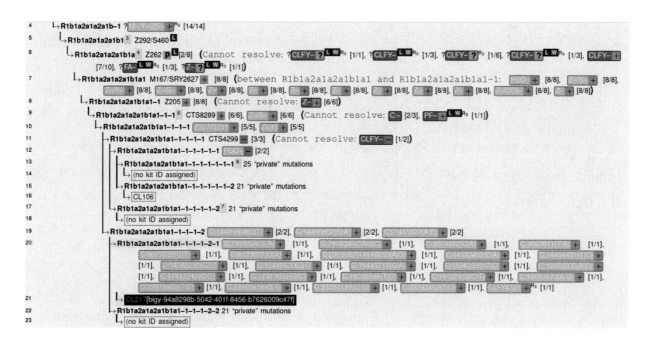

R1b1a2a1a2a1b1a1-1

Links:
------
[1] https://www.clarifydna.com/users/jimrader
[2] https://www.clarifydna.com/contact

R-P312 P312 • S116 • PF6547 Z1904 • CTS12684 • PF6548
  R-DF27 DF27 • S250
    R-Z195 Z196 • S355 Z195 • S227
     R-Z198 Z198 • S228
        R-Z292 ZS312 • Y964 • M7953 • Y964.2 • M7953.2 • Y964.1
                • M7953.1   Z292 • S460
      R-Z262 Z200 • S361 Z691 Z263 Z267 V3130 • Z689
        • CTS4716 Z262 Z688 M167 • SRY2627 FGC11216 • Z265
        • PF1185 Z201 • S457 Z204 Z199 • S234 Z269
          R-Z202 Z266 Z264 Z205 Z202 Z203
            R-Z206 Z206 • S235 Z690
              R-Z208 Z693 Z208 • S362 Z207 • S251 CTS8289
              R-Y15783 Y15784 Y15646 Y15783
I'm trying to understand the process at ISOGG.
    1. would you be the one to edit Rad Banks "Experimental tree" to add the four
new SNPs below M167/SRY2627?

---

2. and I see FTDNA has other SMPs in the same area
3. how would they be numbered (long name R1b1a2a1a2a1b1a1)?

R-P312 P312 • S116 • PF6547 Z1904 • CTS12684 • PF6548
  R-DF27 DF27 • S250
    R-Z195 Z196 • S355 Z195 • S227
      R-Z198 Z198 • S228
        R-Z292 ZS312 • Y964 • M7953 • Y964.2 • M7953.2 • Y964.1
          • M7953.1   Z292 • S460
        R-Z262 Z200 • S361 Z691 Z263 Z267 V3130 • Z689
          • CTS4716 Z262 Z688 M167 • SRY2627 FGC11216 • Z265
          • PF1185 Z201 • S457 Z204 Z199 • S234 Z269
                R1b1a2a1a2a1b1a1
      R-Z202 Z266 Z264 Z205 Z202 Z203
        R-Z206 Z206 • S235 Z690
          R-Z208 Z693 Z208 • S362 Z207 • S251 CTS8289
            R-Y15783 Y15784 Y15646 Y15783

Novel SNPs (40)					Download .CSV
• Best qual (17)	• Acceptable qual (7)	Ambiguous qual (16)	Low qual (0)	One reading! (0)	• INDELs (0)

		REF		ALT	
ChrY: 7905834	YFS450774	G	to	A	Q:100
ChrY: 8374882	YFS450775	C	to	A	Q:100
ChrY: 9123622	YFS450778	G	to	A	Q:100
ChrY: 13871122	YFS450784	G	to	A	Q:100
ChrY: 14148053	YFS450786	A	to	G	Q:100
ChrY: 14330402	YFS450787	C	to	A	Q:100
ChrY: 14488481	YFS450790	T	to	C	Q:100
ChrY: 14766231	YFS450791	G	to	C	Q:100
ChrY: 15449717	YFS450793	G	to	A	Q:100 G
ChrY: 15700574	YFS450796	C	to	T	Q:100
ChrY: 15942532	YFS450797	C	to	G	Q:100
ChrY: 17417247	YFS450806	C	to	G	Q:100
ChrY: 17517692	YFS450807	G	to	C	Q:100
ChrY: 18122933	YFS450812	C	to	T	Q:100
ChrY: 18693670	YFS450817	A	to	G	Q:100
ChrY: 23353191	YFS450831	C	to	G	Q:100
ChrY: 23360522	YFS450832	A	to	G	Q:100

Chapter 10 - - A progress report on using big Y results.

At this point all four of my test-taker (Harry, Alton, Earl and myself) have Big-Y results run through family tree DNA. I have also loaded the BAM files to Yfull.

A year and a half ago I took the big Y test myself. It was recommended you hire external consultants to analyze the results. I selected all three of the consultants and had them do their analysis. After I had completed that I published the book "Y-DNA Jim Rader's multiple tests to understand his Y-DNA" (Dec 2015) at that point my understanding of what I had completed was very limited.

In the summer of 2016, I saw the Irish DNA work on YouTube, with snips and STRs. They had added the SNP analysis to bring archaeological timeframe ancient history into the current genealogical timeframe. To do that they showed that you needed to have a Y111 test and a big Y tests.

So in the fall of 2016, I tried to get the samples from the "Rader" men who had been active in my 2006 Y37 test updated. With more effort, I contacted the three individuals who had been in my 2006 tests and one who is a newly discovered relative.

My Genealogy research shows me I am related to the three men who I am testing as follows:

4. Earl is a 4th cousin twice removed on my Rader line

5. Alton is a 4th cousin twice removed on my Rader line

6. Harry is a 2nd cousin once removed on my Rader line

   And a 4th cousin twice removed on my Bauer line

   And a 4th cousin twice removed on my Andes line

When the annual FTDNA sale began with coupons, I tried to get tests taken. By the end of December, when the sale ended, I had completed tests for all three of those men. I also added an Autosomal test to each.

The table below entitled "compare novel variants all four 28Jan17" displays a lot of details obtained from that study.

The left column entitled Yseq with FGC names are the locations which I had tested with Sanger test methods to verify that the results from the big Y were accurate.

The second column is the name that these particular locations are known by, at some companies.

The next section is the results of my big Y tests. The numbers are the physical location being tested.

The section which is not highlighted his Harry's test results followed by the section with Earl's test results and the final section on the right-hand side is Alton's results

I aligned them so that the places where there are blank spaces are locations where we do not match! You will see that we do not match an equal amount or in the same places. Jim has 23 unique variants; Harry has 16 variants, Earl has 28, and Alton 22.

compare novel variants all 4 28jan17

Yseq verified			Jim 23			Harry 16		Earl 28		Alton 22		
		x	Position	Re	Gei	Position	Gei	Position		Genotype		
								13599179	G			
FGC35933	YFS450786	x	14148053	A	G			14148053	G			
FGC35934	YFS450787	x	14330402	C	A			14330402	A	14330402	A	High
FGC35936	YFS450790	x	14488481	T	C					14488481	C	High
FGC35937	YFS450791	x	14766231	G	C	14766231	C	14766231	C	14766231	C	High
FGC35938	YFS450792	x	15449620	C	T	15449620	T	15449620	T	15449620	T	High
FGC35941	YFS450797	x	15942532	C	G	15942532	G	15942532	G	15942532	G	High
FGC35942	YFS450805	x	16856500	A	G	16856500	G	16856500	G	16856500	G	High
FGC35943	YFS450806	x	17417247	C	G	17417247	G	17417247	G	17417247	G	High
FGC35944	YFS450807	x	17517692	G	C	17517692	C	17517692	C	17517692	C	High
								17887770	T			
FGC35945	YFS450810	x	18032596	T	C	18032596	C	18032596	C	18032596	C	High
FGC35946	YFS450812	x	18122933	C	T	18122933	T	18122933	T	18122933	T	High
								18396980	A			
FGC35947	YFS450817	x	18693670	A	G	18693670	G	18693670	G			
		x	22239121	C	A			22239121	A	22239121	A	High
FGC35948		x	22271952	T	C	22271952	C	22271952	C	22271952	C	High
										22437423	T	High
FGC35949		x	22465458	G	T	22465458	T	22465458	T	22465458	T	High
								22471883	T			
								22471895	G			
								22471901	G			
FGC35950	YFS450831	x	23353191	C	G	23353191	G	23353191	G	23353191	G	High
FGC35952	YFS92463	x	23392463	T	G	23392463	G	23392463	G	23392463	G	High
FGC35929	YFS450774	x	7905834	G	A	7905834	A	7905834	A	7905834	A	High
FGC35930	YFS450775	x	8374882	C	A			8374882	A	8374882	A	High
	YFS450776	x	9027611	T	A			9027611	A	9027611	A	High
	YFS450777	x	9027637	C	T			9027637	T	9027637	T	High
FGC35931	YFS450778	x	9123622	G	A	9123622	A	9123622	A	9123622	A	High
		x	9378241	C	T	9378241		9378241	T	9378241	T	High

As results are received there is doubt on whether the NGS test made errors. So I submitted a sample to a different lab for a test of those SNPs found.

YSEQ Allele Results

SampleID	Marker	Chr	Start+	End	A	llele
89	FGC35929	ChrY	7905834	7905834	A	+
89	FGC35930	ChrY	8374882	8374882	A	+
89	FGC35933	ChrY	14148053	14148053	G	+
89	FGC35934	ChrY	14330402	14330402	A	+
89	M2685	ChrY	14330570	14330570	A	+
89	FGC35936	ChrY	14488481	14488481	C	+
89	FGC35937	ChrY	14766231	14766231	C	+
89	FGC35938	ChrY	15449620	15449620	T	+
89	FGC35938	ChrY	15449620	15449620	T	+
89	FGC35939	ChrY	15449717	15449717	A	+
89	FGC35942	ChrY	16856500	16856500	G	+
89	FGC35943	ChrY	17417247	17417247	G	+
89	CTS7517	ChrY	17517587	17517587	A	+
89	M5722	ChrY	17517587	17517587	A	+
89	FGC35944	ChrY	17517692	17517692	C	+
89	FGC35945	ChrY	18032596	18032596	C	+
89	FGC35946	ChrY	18122933	18122933	T	+
89	FGC35947	ChrY	18693670	18693670	G	+
89	FGC35950	ChrY	23353191	23353191	G	+
89	FGC35952	ChrY	23392463	23392463	G	+

As a test is completed, I also submit the tests for analysis to the same three companies we used before. The Russian company Yfull seems to be the best at putting trees together. It is now the end of January 2016, and the forth of those tests is in the analysis.

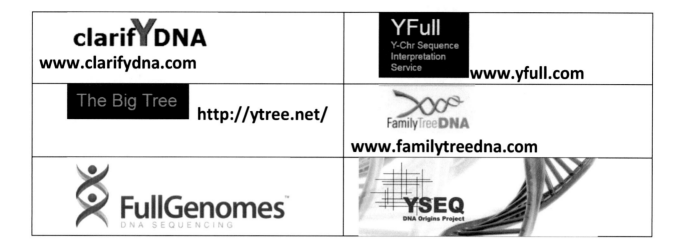

ClarifyDNA results with just two BigY report

I ran only two through ClarifYdna because they are working with older trees which they do not have control of so they don't add the new branches suggested by the results.

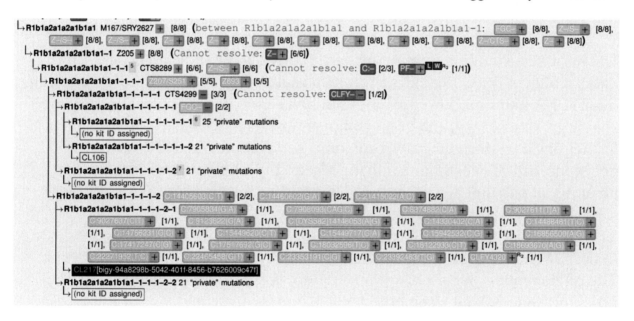

An experimental computer-generated Y-chromosomal phylogeny, supplementing the ISOGG tree with FGC, BigY, and Chromo2 results1 14th May 2015

This report features Y-DNA markers which could identify the test subject or his immediate patrilineal relatives.

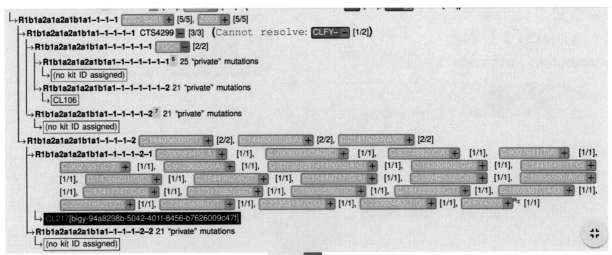

My Haplogroup the old way = R1b1a2a1a2a1b1a1-1-1-1-2-1

Please keep this mind when deciding whether to redistribute this report. The content of this report was produced by a computer algorithm. This report will be re-generated as more information becomes available.

The pilot-scale implementation of clarifYDNA's algorithm can process a dataset of over 4000 BigY kits (over 400 real and 3600 simulated) in one run. clarifYDNA's automation capabilities analyze large Y-SNP data sets with great speed, great accuracy, and great comprehensiveness. These facets are critical for helping a testing company's customers make informed SNP-ordering decisions; uniting customers and/or research participants with their most meaningful patrilineal matches; and, overall, scientific progress, customer satisfaction, and further growth.

ClarifYDNA's software is the key to realizing the "Y Tree" in "Family Tree." The phylogenetic algorithm employed here was initially developed in June 2013 for Geno 2.0 data; see http://ytree.morleydna.com/ experimental-phylogeny for similar reports (from an earlier version of the phylogenetic algorithm) leveraging public Geno 2.0 data. While this report represents a large advance over existing Y-DNA trees, please treat some aspects of this report as experimental and preliminary; some enhancements specific to next-generation sequencing have not been exhaustively tested, and there are several discrepancies over the definitions of high-level SNPs.

Terminal-level mutations marked as recurrent (R) or preceded by a question mark should be taken with some caution clarifYDNA currently processes Y-DNA information from four different datasets. Three of these draw their data from direct-to-consumer genetic genealogy products (BigY, FGC Y Elite, and Chromo2 results). The fourth draws its data from the 1000 Genomes Project [8]. Owing to differences in these datasets' coverage levels, each dataset is processed separately. This results in four phylogenies — one implied by each of the four datasets. Your report consists of an excerpt from up to four trees. One of these trees — your primary tree — is based on your submitted data and other results from that dataset. For example, if you ordered an analysis of your BigY data, then you will receive a clarifYDNA report with the BigY-implied tree as your primary tree.

Reports will also include — subject to data availability — excerpts from the other three trees. These auxiliary trees will show roughly the same genetic neighborhood as in your primary tree. Your kit's results (positive calls, and some of the relevant nonpositive calls) are incorporated into each tree excerpt.

This will help show the positioning of your kit on each of the auxiliary trees. Work is underway to automate the production of a composite tree.

If you found this service useful, please help us spread awareness. Or consider making a donation to offset the months of research and development that went into it.
This report may not be used for commercial purposes.

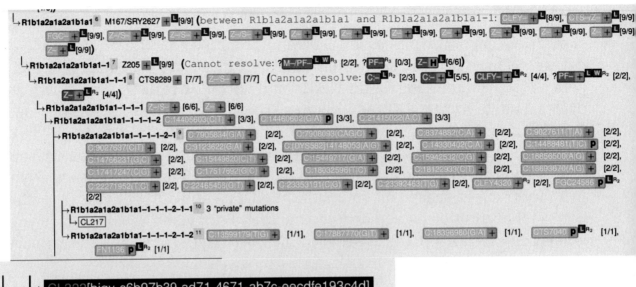

↳ R1b1a2a1a2a1b1a1 [6] M167/SRY2627 ⊞ L[9/9] (between R1b1a2a1a2a1b1a1 and R1b1a2a1a2a1b1a1-1: CLFY– ⊞ L[8/9], CTS–Z– ⊞ L[9/9] FGC– ⊞ L[9/9], Z–/S– ⊞ L[9/9], Z–/S– ⊞ L[9/9], Z–/S– ⊞ L[9/9], Z– ⊞ L[9/9], Z– ⊞ L[9/9], Z– ⊞ L[9/9], Z– ⊞ L[9/9], Z– ⊞ L[9/9], Z– ⊞ L[9/9])

↳ R1b1a2a1a2a1b1a1–1 [7] Z205 ⊞ [9/9] (Cannot resolve: ?M–/PF– L W[R3] [2/2], ?PF– [R3] [0/3], Z– H L[6/6])

↳ R1b1a2a1a2a1b1a1–1–1 [8] CTS8289 ⊞ [7/7], Z–/S– ⊞ [7/7] (Cannot resolve: C– L[R2] [2/3], C– L[R2] [5/5], CLFY– ⊞ L[R2] [4/4], ?PF– L W[R2] [2/2], Z– ⊞ L[R2] [4/4])

↳ R1b1a2a1a2a1b1a1–1–1–1 Z–/S– ⊞ [6/6], Z– ⊞ [6/6]

↳ R1b1a2a1a2a1b1a1–1–1–1–2 C:14405603(C|T) ⊞ [3/3], C:14460602(G|A) p [3/3], C:21415022(A|C) ⊞ [3/3]

↳ R1b1a2a1a2a1b1a1–1–1–1–2–1 [9] C:7905834(G|A) ⊞ [2/2], C:7908093(CAG|C) ⊞ [2/2], C:8374882(C|A) ⊞ [2/2], C:9027611(T|A) ⊞ [2/2], C:9027837(C|T) ⊞ [2/2], C:9123622(G|A) ⊞ [2/2], C:[DYS582]14148053(A|G) ⊞ [2/2], C:14330402(C|A) ⊞ [2/2], C:14488481(T|C) p [2/2], C:14766231(G|C) ⊞ [2/2], C:15449820(C|T) ⊞ [2/2], C:15449717(G|A) ⊞ [2/2], C:15942532(C|G) ⊞ [2/2], C:16856500(A|G) ⊞ [2/2], C:17417247(C|G) ⊞ [2/2], C:17517692(G|C) ⊞ [2/2], C:18032596(T|C) ⊞ [2/2], C:18122933(C|T) ⊞ [2/2], C:18693670(A|G) ⊞ [2/2], C:22271952(T|C) ⊞ [2/2], C:22465458(G|T) ⊞ [2/2], C:23353191(C|G) ⊞ [2/2], C:23392463(T|G) ⊞ [2/2], CLFY4320 ⊞[R2] [2/2], FGC24586 p L[2/2]

↳ R1b1a2a1a2a1b1a1–1–1–1–2–1–1 [10] 3 "private" mutations
  ↳ CL217

↳ R1b1a2a1a2a1b1a1–1–1–1–2–1–2 [11] C:13599179(T|G) ⊞ [1/1], C:17887770(G|T) ⊞ [1/1], C:18396980(G|A) ⊞ [1/1], CTS7040 p L[R2] [1/1], FN1136 p L[R2] [1/1]

↳ CL322[bigy-c6b97b39-ad71-4671-ab7c-eecdfe193c4d]

↳ **R1b1a2a1a2a1b1a1–1–1–1–2–2** [12] 22 "private" mutations
  ↳ (no kit ID assigned)

Symbol Class	Symbol	Description
SNPs and INDELs	M343	Mutations with phylogenetic positions known to ISOGG [1]. If on a branch represented in the dataset: then the available Y-DNA results are consistent with ISOGG's positioning [1].
	CTS3368	Mutation unplaced in ISOGG's tree – placement has been proposed by the phylogenetic algorithm.
	L440	Expected by ISOGG to be at this location, but the data suggests otherwise. Possible causes: (i) mutation not consistently covered in tests; (ii) cannot pinpoint position in clade because sibling clades lack representation in the dataset.
	CTS2526 ^{R_x}	The mutation is recurrent – it has been also placed elsewhere in this tree. Recurrent does not necessarily mean erratic. The subscript x indicates how many instances of the mutation are observed in this tree. Highly recurrent markers are omitted from report.
	PAGE65 [b]	ISOGG [1] has identified this instance as a back-mutation.
	C:12345(A|T)	A mutation that hasn't been named. This example represents a mutation from A to T at the 12345 position of the Y chromosome.
	C:22293662(C|G) [c]	A mutation with this superscript is suspiciously close to another novel variant at this level. These nearby mutations could be part of the same mutation, or the manifestations of an alignment or mapping error.
	?FGC603653	This mutation may actually lie higher up on the tree, only appearing to be recurrent due to spotty coverage.
	Y201	This mutation's phylogenetic placement is uncertain, most likely due to inconsistent or conflicting results.

Subject kit's results	+	Report subject is positive for this mutation.
	p	Report subject has a weak positive for this mutation.
	−	Report subject is negative for this mutation.
	H	Report subject is heterozygous at this position (or nearby).
Kits	CL55	Kits that have been assigned a clarifY ID.
Clades	**A1**	A clade (hierarchical level) in the human Y tree.
	$-n$	Represents a subclade new to ISOGG's tree, proposed by the author's phylogenetic algorithm. For example, **R1b1–1** is proposed to be downstream of **R1b1** (and upstream of **R1b1a**). Some of these new subclades, if terminal, may have already been deemed "private" by ISOGG.
	$\overset{?}{\rightarrow}$	This branch does not feature any BigY-tested kits, and consequently its position in this phylogeny is unconfirmed.
	A1 [4]	This clade has a footnote. Footnotes usually describe automatically detected discrepancies in the tree.
Coverage statistics	$[x/y]$	y is the number of kits supposedly downstream of this mutation which have clear results for this mutation, and x is the number of these kits that are positive for the mutation.

R-Y15783 Y15783 * Y15646 * FGC35927/Y15784 formed 3200 ybp, TMRCA 2900 ybp info

- id:YF08017 new
- id:YF07939 new
- id:YF06136
- id:YF03497 USA [US-CA]
- id:NA19777 MXL

Earl
Harry
?
Jim
?

By James Lee Rader

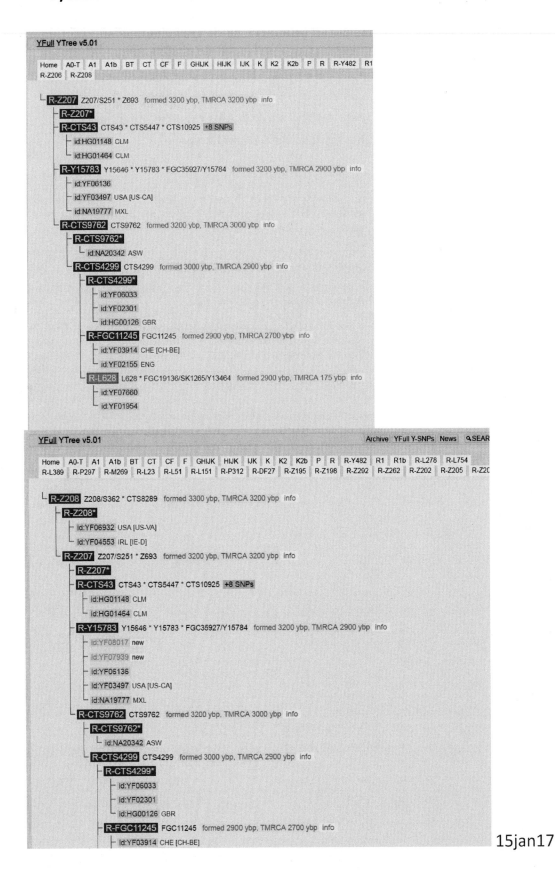

15jan17

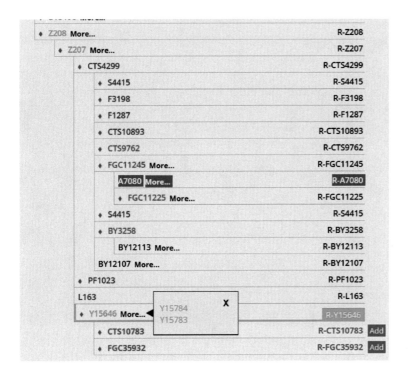

Chapter 11 - -FTDNA reports for Big-Y

Y-DNA Reports

Jim

Earl

Match Name	Shared Novel Variants	Known SNP Difference	Non-Matching Known SNPs	Matching SNPs	Match Date
	Show All ▾	Show All ▾			
👤 Alton Rader Managed by James Lee Rader ✉ 💬	36	0		25,995	1/18/2017
👤 Harry Wayne Rader Managed by James Lee Rader ✉ 💬	31	0		25,115	12/22/2016
James Lee Rader ✉ 💬	22	0		26,659	12/22/2016
👤 Adam J. Peiper ✉ 💬	19	0		26,247	12/22/2016
E. Randol Schoenberg ✉ 💬	1	0		26,537	12/22/2016
👤 Patrick Redgate (ADOPTED) ✉ 💬	0	0		26,447	12/22/2016
👤 Mr. Patrick Wangermez ✉ 💬	20	1	CTS4299	26,639	12/22/2016
👤 Donald Edward Guthrie ✉ 💬	19	1	CTS4299	26,587	12/22/2016
Mr. Jaime Enrique Conde Matos ✉ 💬	17	1	CTS4299	26,412	12/22/2016
👤 Xavier Devaugermé ✉ 💬	17	1	CTS4299	26,309	12/22/2016
👤 Robert Luther Hodges ✉ 💬	15	1	CTS4299	26,478	3/9/2017
👤 Mr. Terry Lee Palmer Sr. ✉ 💬	4	1	CTS10783	26,658	12/22/2016
👤 John Thomas Parish ✉ 💬	1	1	CTS4299	26,348	12/22/2016
Mr. Patrick David Berge ✉ 💬	1	1	CTS4299	26,610	12/22/2016
Mr. Robert Kiefer ✉ 💬	1	1	F2333	26,497	12/22/2016
👤 Mr. Thierry Wangermez ✉ 💬	1	1	CTS4299	26,245	12/22/2016

Alton

Family Finder reports – Rader relatives

Name	Match Date	Relationship Range	Shared Centimorgans	Longest Block	X-Match	Linked Relationship	Ancestral Surnames	+
Earl Francis Rader Managed by James Lee Rader	12/28/2016	2nd Cousin - 4th Cousin	85	32				
Ms. Laurie Beaton	12/28/2016	2nd Cousin - 4th Cousin	62	37			Ashford / Bursk (Manchester, England) / Burstein (England, Russia) / Cooper /	
Manda Colburn	12/28/2016	3rd Cousin - 5th Cousin	53	15			SARA / Mothe / M Agne / Exelin / LEA / ELIZABET / Elizabeth unverifie / Sara /	
Pamela McCunn	12/28/2016	2nd Cousin - 4th Cousin	48	17			Hanna / Sall / Addr / Anna Mari / Margare / France / Suzann / Mar / Agne / Catherin /	
Mr. DANIEL Mc NEIL Witt	12/28/2016	2nd Cousin - 4th Cousin	48	23			(Estella) / Addi / Bromley / Campeau / Charles / Chauvin / Cunningham / Daniels	
M.C. Southerland	12/28/2016	4th Cousin - Remote Cousin	39	11			Aschenbach / Alexander (Greene and Cocke Counties, Tennessee, Virginia,	
Mr. Thomas Arthur Bean Jr.	12/28/2016	5th Cousin - Remote Cousin	39	8			Narciss / An / Ros / Barbar / Abigai / Elizabet / Margare / Elisabet / Avey /	
Susan Renee Steveson	12/28/2016	5th Cousin - Remote Cousin	39	8			/ Angel / Becker / Baker / BLAIR / Bingham / BRICKER / Bricker / Bricker (Iowa)	
Harry Wayne Rader Managed by James Lee Rader	12/28/2016	5th Cousin - Remote Cousin	38	10				

Alton Rader Managed...
Ethnic Makeup

∧ European — 93%
British Isles — 76%
Southeast Europe — 11%
Finland — 6%

∧ Jewish Diaspora — 6%
Sephardic — 6%

∧ Trace Results
Ashkenazi — <2%

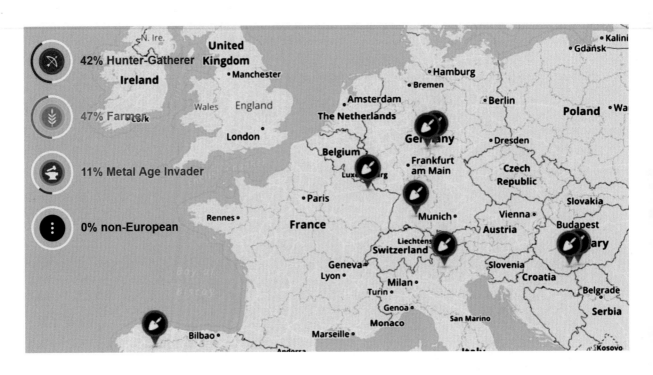

Y-DNA Reports

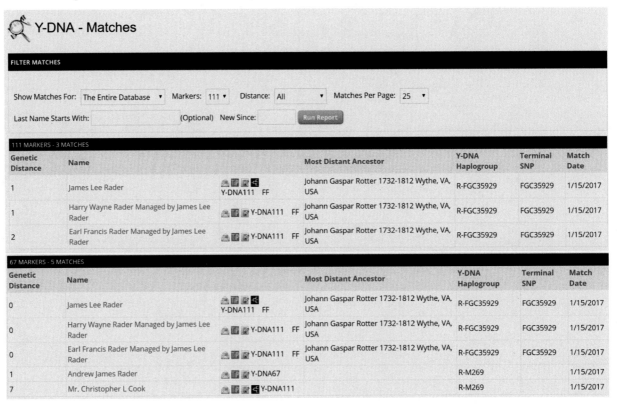

Genetic Distance	Name		Most Distant Ancestor	Y-DNA Haplogroup	Terminal SNP	Match Date
0	Michael David Rader	Y-DNA37	John Rader b. 1796	R-M269		1/4/2017
0	Mr. Robert Merl Rader Sr.	Y-DNA37 FF		R-M269		1/4/2017
0	Alton Clyde Rader	Y-DNA37		R-M269		1/4/2017
0	James Lee Rader	Y-DNA111 FF	Johann Gaspar Rotter 1732-1812 Wythe, VA, USA	R-FGC35929	FGC35929	1/4/2017
0	Harry Wayne Rader Managed by James Lee Rader	Y-DNA111 FF	Johann Gaspar Rotter 1732-1812 Wythe, VA, USA	R-FGC35929	FGC35929	1/4/2017
0	Earl Francis Rader Managed by James Lee Rader	Y-DNA111 FF	Johann Gaspar Rotter 1732-1812 Wythe, VA, USA	R-FGC35929	FGC35929	1/4/2017
1	Mr. Robert J Myers	Y-DNA37 FF		R-M269		1/4/2017
1	Andrew James Rader	Y-DNA67		R-M269		1/4/2017
4	Thomas Frederick Johnson	Y-DNA111	John N. Johnson b 1823 d 1912	R-M269		1/4/2017
4	Pamela Stone Eagleson	Y-DNA37	Francis Stone m.1702 Richmond County VA	R-M269		1/4/2017
4	Philip Stone	Y-DNA111 FF		R-M269		1/4/2017
4	David C. Stone	Y-DNA111 FF	Charles Isaac McLain Stone	R-M269		1/4/2017
4	Jervis		T008 Ricketts, William d 1700 Jamaica	R-M269		1/4/2017
4	Mr. Bassett			R-M269		1/4/2017
4	Lapworth		William Lapworth, 29-7-1773 coventry, england	R-M269		1/4/2017

37 MARKERS - 15 MATCHES

Harry

Match Name	Shared Novel Variants	Known SNP Difference	Non-Matching Known SNPs	Matching SNPs	Match Date
Name Search	Show All ▾	Show All ▾	SNP Name Search		Match Date S
Earl Francis Rader Managed by James Lee Rader	36	0		25,995	1/18/2017
Harry Wayne Rader Managed by James Lee Rader	30	0		24,973	1/18/2017
James Lee Rader	21	0		26,065	1/18/2017
Adam J. Peiper	17	0		25,876	1/18/2017
E. Randol Schoenberg	2	0		26,027	1/18/2017
Patrick Redgate (ADOPTED)	0	0		25,967	1/18/2017
Daniel Michael Parrish	17	1	CTS4299	25,947	1/26/2017
Luc Vangermeersch	17	1	CTS4299	26,006	1/18/2017
Mr. Henri Wangermez	17	1	CTS4299	26,067	1/18/2017
Mr. Patrick Wangermez	17	1	CTS4299	26,053	1/18/2017
Xavier Devaugermé	17	1	CTS4299	25,920	1/18/2017
Donald Edward Guthrie	16	1	CTS4299	26,040	1/18/2017
Mr. Jaime Enrique Conde Matos	16	1	CTS4299	25,962	1/18/2017

Chapter 12 - - YFull reports

Their home page states the following! https://yfull.com/

We'll solve your Y-Chromosome Puzzle!

Technical progress never has a standstill, and technologies in DNA sequence analysis of next generation have developed so promptly recently, that volumes of received initial information exceed tens gigabytes, and complexity of its processing demands professional interpretation tools. We'll analyze your NextGen Y-Chr RAW data

HAPLOGROUP AND Y-SNPS
Unlock your terminal haplogroup and known Y-SNPs (105k+)

NOVEL VARIANTS
Discover unique Y-SNPs found only in your sample

Y-STRS
Report on all the STRs (500+) extracted from Y-Chromosome."

NextGen Sequence Interpretation $49

— Analysis and comparing your NextGen Y-Chr sequencing data
— YReport (known SNPs, STRs, novel SNPs, STR and SNP matches)
— All newly discovered SNPs will be added to the Y-series
— Your data can be used anonymously in the YFull research projects
— Actual terminal SNPs and positioning on the YTree
— You may use group membership (haplogroups, geographical or family)
— Access to the raw data at any time (free storage)
— Free regular updates!
 * Technical requirements: BAM file; depth of coverage min 15X; read length min 100 bp
 * No subscription fee! No prepayments. MtDNA results optionally

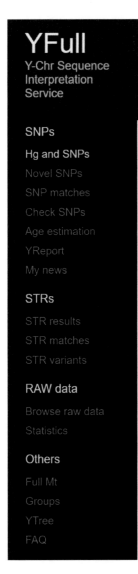

YFull
Y-Chr Sequence
Interpretation
Service

SNPs

Hg and SNPs

Novel SNPs

SNP matches

Check SNPs

Age estimation

YReport

My news

STRs

STR results

STR matches

STR variants

RAW data

Browse raw data

Statistics

Others

Full Mt

Groups

YTree

FAQ

Yfull results based on YFull YTree v5.03 at 09 April 2017

R-FGC35930

Four men took the tests, and their results were transferred to Y full which is a company in Russia. They took some time to do their analysis of the BAM files as they are called. The results came in over several months time, but these BAM files were also sent over two months time.

As you look at the various reports, you must know of these four numbers. These are the numbers assigned by Y Full to designate the four kits he received.

YF03497 Jim
YF07939 Harry
YF08017 Earl
YF08525 Alton

They are listed in the order they were submitted.

Family Tree DNA specifically lists this test is an experimental test and instructs the user to hire an outside company to analyze the results. In the previous chapter, you seen the reports generated by family tree DNA.

Other information from Big-Y

Does the family have any unique mutations?
It is confusing to understand the mutations versus snips. A unique mutation is something one individual has and the no other one has shown up yet. As soon as you get two or more people who share a unique mutation, it is no longer unique mutation, but it becomes a snip.

What other groups share unique mutations
As time goes by more people's tests are analyze and placed on various trees. Y full updates are tree about monthly so you should go back each month to see who matches you.

When and where did splits occur
The question of where the split occurred is determined by where the people lived which we don't know. The anthropologist's community is busy

analyzing their samples collected over the years that the sciences existed and using the same new tests there trying to get more specific snips that apply to their particular samples which will give us a better representation of which people lived where or more specifically died where.

When did the split occur is the question and that is answered by using the average time between snips. So if you look at this case where we are about 3000 years ago on that is the Z207 and divided by these 20 snips we have showing up on these tests. 3000 years divided by 20 snips if you hundred and 150 years per snip.

Put the results together
so one would expect on average the most recent snips that were showing occurred 150 years ago.

The graphics on these pages will show you the reports generated by y full as they received the samples and analyze them. We have the snip mutations and the age estimation created by Y full. The STR differences are coming in much slower as Y full develops a methodology to report them. Family Tree DNA doesn't even acknowledge there are strs reported in the big Y test.

110 to 170 years between new SNPs

When you look at the following graphic, there are two parts of the graphic the top half shows which samples are in the particular snip. The bottom half shows the calculation that comes up with when the snips occurred. So if you refer to the start of this chapter, you will remember there are four numbers listed there, but in this graphic, you only see one number the YF03497 which is my identification number. And then there are two other people you'll see him ask Al after one of them meaning that person was from Mexico I believe and the other one is from another test. So after they received my results and analyze them, they came up with their best guess which was the might tests came from 2884 years before the present time

Age with 1 Big-Y test

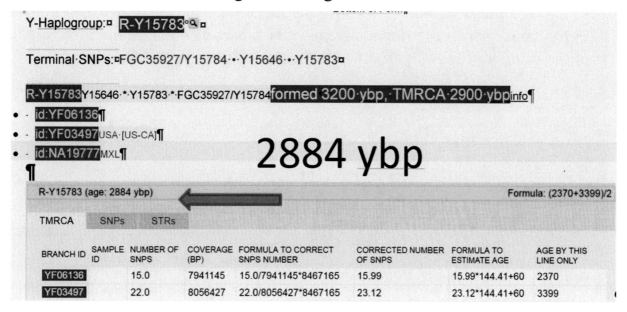

They had the three tests then I now have decided that all three are in a new snip and that the other two people being included are not part of this calculation there in the difference snip. You'll also see their calculation is born it must closer in time to under 400 years

Age with 3 Big-Y tests

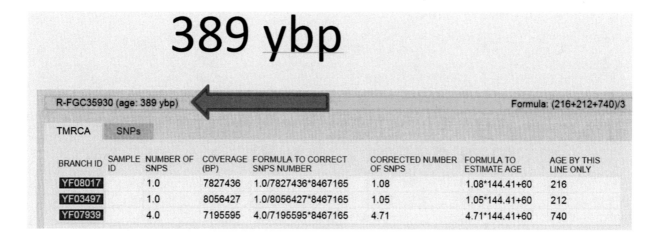

And the forth test added gives them an even closer estimation of under 300 years you'll also notice that my test and Alton's tests both show a distance of zero snips

where Harry's test shows four steps in Earl's test shows one snip. And if you refer to the tree you remember that Harry is upstream from me, so they are saying Harry has four snips off of the average and I, a person who is in the same branch of the same tree and further down the tree has fewer snips as in zero snips.

Age with 4 Big-Y tests

269 ybp

R-FGC35930 (age: 269 ybp)   Formula: (60+740+216+60)/4

TMRCA   SNPs

BRANCH ID	SAMPLE ID	NUMBER OF SNPS	COVERAGE (BP)	FORMULA TO CORRECT SNPS NUMBER	CORRECTED NUMBER OF SNPS	FORMULA TO ESTIMATE AGE	AGE BY THIS LINE ONLY
YF03497		0.0	8056427	0.0/8056427*8467165	0.0	0.0*144.41+60	60
YF07939		4.0	7195595	4.0/7195595*8467165	4.71	4.71*144.41+60	740
YF08017		1.0	7827436	1.0/7827436*8467165	1.08	1.08*144.41+60	216
YF08525		0.0	7588394	0.0/7588394*8467165	0.0	0.0*144.41+60	60

Relating this back to my genealogy

Casper was born 1732
- 2017
285

Estimated age 269 ybp

This report from Y full is showing the known snips for this group of people at the point in time where the Y tree is in version 5.03. As we get more people testing and more time to analyze we will see these reports will change on Yfull what was a unique snip will move from that page to this page. At that point according to wife all it's not a unique snip anymore it is an actual shared snip

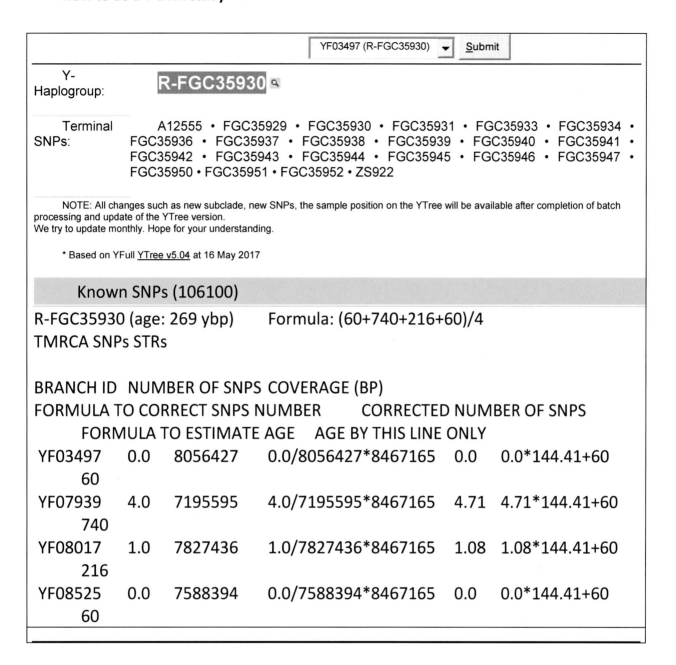

The following is part of the tree as understood by Y full in version 5.03. I've included the part of the tree from what we were known before this test which was Z207. Just below that, you will see R-Y 15783 which is the place in the tree where family tree DNA is presently showing us as being located. There is another new level called R-FGC35930 which is the more detailed and newer snip discovered and Y full's analysis. And you will see that all four of us who've tested with this set of the tests are grouped together in her unique location on the tree. And listed in reverse order one on the top is the most recent one reported.

The known snips report above was for Jim Rader this is the same report, but it is for Harry Rader. They're both from the same tree results, but you'll see the snips reported are shown in a different order. They both show the quantity of known snips is 103862. So that's over 100,000 snips.

So the next two pages contain the same two reports from the same version of the tree for the other two men who tested.

The following four sections of text are the snips shared according to Y full.

MRCA branch,TMRCA (ybp),Country of origin,ID,Terminal Hg,Shared SNPs,Assumed shared SNPs
R-FGC35930,275 (125-500),United States,YF08525,R-FGC35930,A12555, FGC35929, FGC35930, FGC35931, FGC35933, FGC35934, FGC35937, FGC35938, FGC35939, FGC35940, FGC35941, FGC35942, FGC35943, FGC35944, FGC35946, FGC35947, FGC35950, FGC35952, FGC35936, FGC35945, FGC35951, YSC0000144, YSC0000150, A125, CTS12254, CTS12427/PF1324, CTS12428, CTS12429, CTS12430, CTS12431, CTS12440, CTS12441/S3833, CTS12442/PF2778, CTS12443, CTS12444, CTS12445, CTS12446, CTS12447, CTS12448, CTS12449/S4803, CTS12450/S4062, CTS12453, CTS12454, CTS12455, CTS12456, CTS12457, CTS12458, CTS12460, CTS12461, CTS12462, PF1325, PF362, PF363, PF380, PF382, PF384, PF392, PF394, PF400, PF408, PF415, PF421, PF422/Z2927, PF431, PF433, PF443, PF519, PF528, PF540, PF620, YSC0000126, YSC0000129, YSC0000130, YSC0000137, YSC0000138, YSC0000143, YSC0000146, PF1156, PF1193;ZS922, A395, YSC0000147, YSC0000152, CTS12252, PF1103, PF385, PF403, PF423, PF432, PF496, PF506, PF536, PF608, PF628, PF6737, PF6813, S13575, S13968, S21353, S3038, YSC0000145, YSC0000148, YSC0000149, YSC0000151, Z2028, PF6724, Z3427, CTS2777, PF1184, PF1356, PF355, PF365, PF381, PF393, PF418, PF520, PF535, S14526, Z6731, PF1191;120

R-FGC35930,275 (125-500),United States,YF08017,R-FGC35930,FGC35929, FGC35930, FGC35931, FGC35933, FGC35934, FGC35937, FGC35938, FGC35939, FGC35940, FGC35941, FGC35942, FGC35943, FGC35944, FGC35946, FGC35947, FGC35950, FGC35952, FGC35936, FGC35945, FGC35951, ZS922, M6494,

YSC0000144, YSC0000147, YSC0000150, A125, CTS12254, CTS12440, CTS12441/S3833, CTS12442/PF2778, CTS12443, CTS12444, CTS12445, CTS12446, CTS12447, CTS12448, CTS12449/S4803, CTS12450/S4062, CTS12453, CTS12454, CTS12455, CTS12456, CTS12457, CTS12458, CTS12460, CTS12461, CTS12462, PF1325, PF362, PF363, PF382, PF384, PF385, PF394, PF400, PF408, PF415, PF421, PF422/Z2927, PF431, PF433, PF519, PF536, PF6813, S13575, S21353, YSC0000126, YSC0000129, YSC0000130, YSC0000137, YSC0000138, YSC0000143, YSC0000146, YSC0000148, YSC0000149, YSC0000151, Z2028, PF1156, PF1193;A12555, A395, YSC0000152, CTS12252, CTS12427/PF1324, CTS12428, CTS12429, CTS12430, CTS12431, PF1103, PF380, PF392, PF423, PF432, PF443, PF496, PF506, PF528, PF608, PF620, PF628, PF6737, S13968, S3038, YSC0000145, PF6724, Z3427, CTS2777, PF1184, PF1356, PF355, PF365, PF381, PF393, PF418, PF520, PF535, S14526, Z6731, PF1191;119,

R-FGC35930,275 (125-500),United States,YF07939,R-FGC35930,A12555, FGC35929, FGC35930, FGC35931, FGC35933, FGC35937, FGC35938, FGC35939, FGC35940, FGC35941, FGC35942, FGC35943, FGC35944, FGC35946, FGC35947, FGC35950, FGC35952, FGC35936, FGC35945, YSC0000144, YSC0000147, YSC0000150, A125, CTS12254, CTS12440, CTS12441/S3833, CTS12442/PF2778, CTS12443, CTS12444, CTS12445, CTS12446, CTS12447, CTS12448, CTS12449/S4803, CTS12450/S4062, CTS12453, CTS12454, CTS12455, CTS12456, CTS12457, CTS12458, CTS12460, CTS12461, CTS12462, PF1325, PF362, PF363, PF382, PF392, PF394, PF400, PF421, PF422/Z2927, PF431, PF433, PF506, PF519, PF628, YSC0000126, YSC0000129, YSC0000130, YSC0000137, YSC0000138, YSC0000143, YSC0000145, YSC0000146, YSC0000148, YSC0000149, YSC0000151, PF1156;FGC35934, FGC35951, ZS922, A395, M6494, YSC0000152, CTS12252, CTS12427/PF1324, CTS12428, CTS12429, CTS12430, CTS12431, PF1103, PF380, PF384, PF385, PF403, PF408, PF415, PF423, PF432, PF443, PF496, PF528, PF536, PF540, PF608, PF620, PF6813, S13575, S13968, S3038, Z2028, PF1193, PF6724, CTS2777, CTS9790, CTS9791, PF1184, PF1356, PF355, PF365, PF381, PF393, PF418, PF520, PF535, S14526, Z6731, PF1191;120

The next four graphics are reports for each one of the four people who tested they show you that each person matches the other three people. They also show you how many shared snips they have

---

85

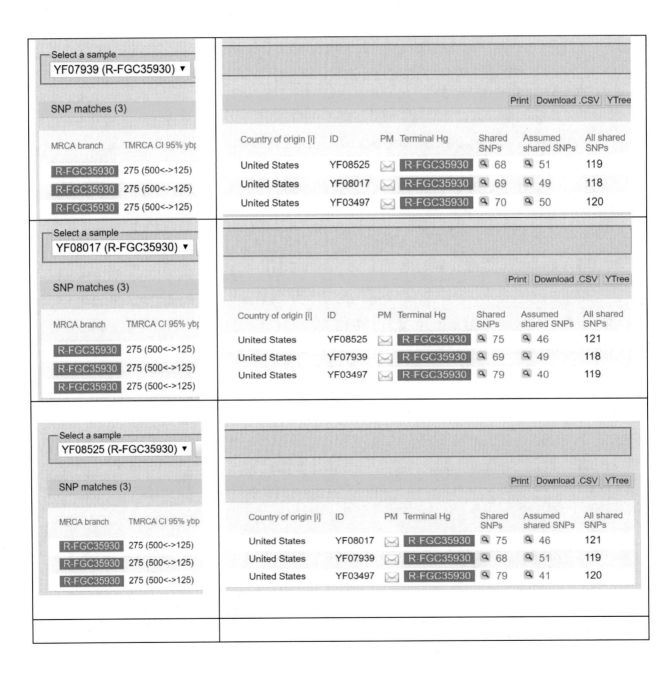

Accuracy of Big-Y test according to YFull

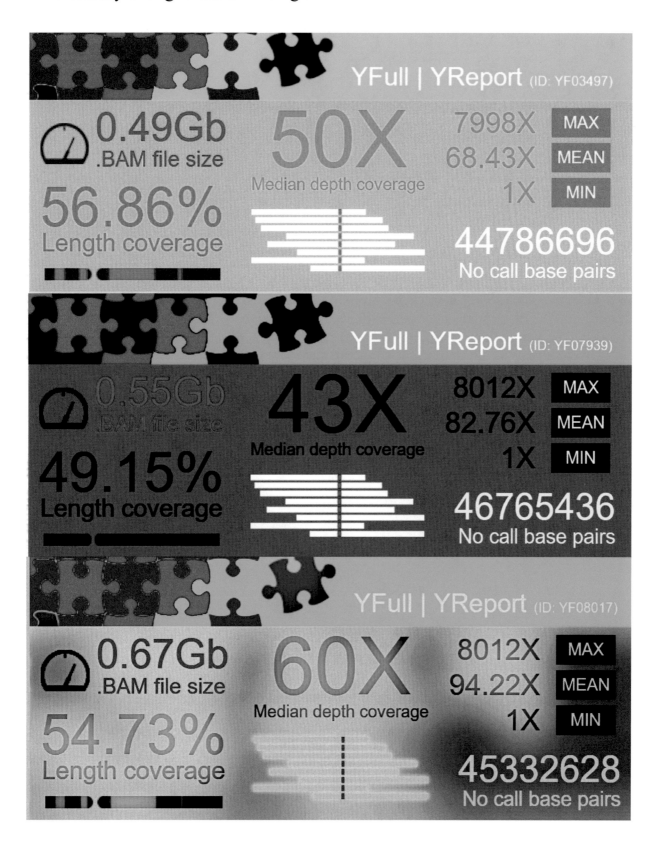

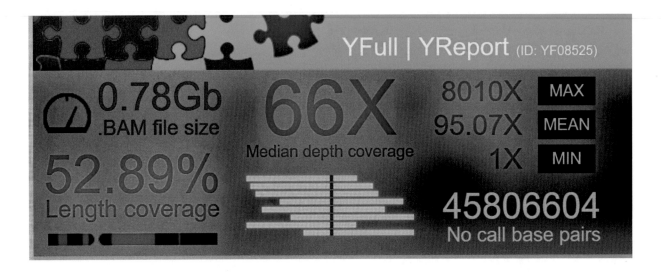

Yfull finishes, and we no longer have unique SNPs

All of the unique SNPs have been converted to shared SNPs

      The Black box contains those SNPs that once were unique but are now shared

      If you look carefully, you will see "info" with a white background clicking on that spot reveals the black box with the additional shared SNPs

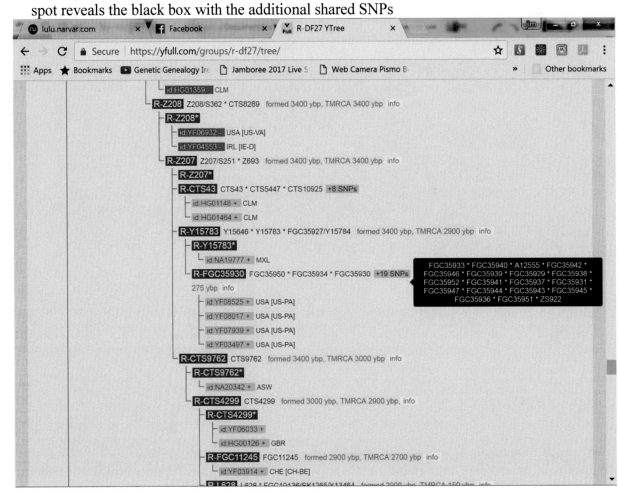

For those of us with vision problems the following is the same tree with the font sizes increased!

R-Z208 **Z208/S362 * CTS8289** formed 3400 ybp, TMRCA 3400 ybp info

- R-Z208*
  - id:YF06932 - USA [US-VA]
  - id:YF04553 - IRL [IE-D]
- R-Z207 Z207/S251 * Z693 formed 3400 ybp, TMRCA 3400 ybp info
  - R-Z207*
  - R-CTS43 CTS43 * CTS5447 * CTS10925+8 SNPs
    - id:HG01148 + CLM
    - id:HG01464 + CLM
  - R-Y15783 Y15646 * Y15783 * FGC35927/Y15784 formed 3400 ybp, TMRCA 2900 ybp info
    - R-Y15783*
      - id:NA19777 + MXL
- R-FGC35930 FGC35950 * FGC35934 * FGC35930+19 SNPs formed 2900 ybp, TMRCA 275 ybp info
  - id:YF08525 + USA [US-PA]
  - id:YF08017 + USA [US-PA]
  - id:YF07939 + USA [US-PA]
  - id:YF03497 + USA [US-PA]

The following box is a supporting window which pops up to allow us to see the calculation used to determine "TMRCA" calculations

BRANCH ID	UMBER OF SNPS	COVERAGE (BP)	FORMULA TO CORRECT SNPS NUMBER	CORRECTED NUMBER OF SNPS	FORMULA TO ESTIMATE AGE	AGE BY THIS LINE ONLY
YF03497	0.0	8056427	0.0/8056427*8467165	0.0	0.0*144.41+60	60
YF07939	4.0	7195595	4.0/7195595*8467165	4.71	4.71*144.41+60	740
YF08017	1.0	7827436	1.0/7827436*8467165	1.08	1.08*144.41+60	216
YF08525	0.0	7588394	0.0/7588394*8467165	0.0	0.0*144.41+60	60

Chapter -13 - How to place your terminal SNP in time and place

Genetic Genealogy

Ancestral Journeys: The Peopling of Europe from the First Venturers to the Vikings (Revised and Updated Edition)Feb 15, 2016, by Jean Manco Paperback $ 11 30 (the paperback is an update from the hardbound so only the paperback)

Review

An interesting account of the peopling of Europe that attempts to integrate archaeology, history, and linguistics with the latest genetic evidence. . . . Recommended. — Choice

The armchair route to uncovering the mysteries of who we really are. — Woman Around Town

Richly illustrated . . . this highly readable volume outlines a new paradigm in European archaeology and pre-history and tackles the central question of the meaning of European identity, genetic and cultural. — ProtoView

New Genetics and Society **ISSN: 1463-6778 (Print) 1469-9915 (Online) Journal homepage:** http://www.tandfonline.com/loi/cngs20

"They want to know where they came from": population genetics, identity, and family genealogy Richard Tutton To cite this article: Richard Tutton (2004) "They want to know where they came from": population genetics, identity, and family genealogy, New Genetics and Society, 23:1, 105-120, DOI: 10.1080/1463677042000189606 To link to this article: http://dx.doi.org/10.1080/1463677042000189606

ABSTRACT This paper discusses the changing relationship between population genetics, family genealogy and identity. It reports on empirical research with participants in a genetic study who anticipated that personal feedback on the analysis of their donated samples would elucidate aspects of their own family genealogies. The paper also documents how geneticists, building on the practices of offering personal feedback to research participants, have developed genetic tests marketed directly to people wishing to trace their ancestry. Some of the social and ethical issues raised by this development in the use of genetic testing are considered.

bell beaker behemoth coming out soon

Eurogenes Blog: The Bell Beaker Behemoth (Olalde et al. 2017 preprint)
eurogenes.blogspot.com/2017/05/the-bell-beaker-behemoth_10.html
May 10, 2017 - Abstract: **Bell Beaker** pottery spread across western and central Europe beginning around 2750 BCE before disappearing between 2200-1800 ...

Eurogenes Blog: Latest on Bell Beaker and Corded Ware
eurogenes.blogspot.com/2017/04/latest-on-bell-beaker-and-corded-ware.html
Apr 3, 2017 - Guess I'll just have to wait for the **Bell Beaker** behemoth. By the way, **Bell Beaker** blogger has a post on the Heyd paper and Razib on the ...

Population genetics: A map of human wanderlust

https://www.nature.com/nature/journal/v538/n7624/full/nature19472.html

- Serena Tucci & Joshua M. Akey *Nature* **538**, 179–180 (13 October 2016) doi:10.1038/nature19472 Published online 21 September 2016

Genetic studies of individuals from geographically diverse human populations provide insights into the dispersal of modern humans across the globe and how geography shaped genomic variation. See Articles p.201 & p.207 & Letter p.238

Subject terms:
Genetics

Genomics

Evolution

A remarkable feature of modern humans is our wanderlust, which the poet Charles Baudelaire famously referred[1] to as *"l'horreur du domicile."* From our evolutionary birthplace in Africa[2], modern humans have migrated to nearly every habitable corner of Earth (Fig. 1), overcoming obstacles such as ice, deserts, oceans, and mountains. The number, timing, and routes of human dispersals out of Africa have implications for understanding our past and how that past influenced contemporary patterns of human genomic variation. Three studies on pages 207, 201 and 238 (Malaspinas *et al.*[3], Mallick *et al.*[4] and

Pagani *et al.*[5]) describe 787 new, high-quality genomes of individuals from geographically diverse populations, providing opportunities to refine and extend current models of historical human migration.

Molecular Anthropology

Wiley: An Introduction to Molecular Anthropology - Mark Stoneking

www.wiley.com/WileyCDA/WileyTitle/productCd-1118061624,subjectCd-AN45.html

An Introduction to **Molecular Anthropology** is an invaluable resource for students studying human evolution, biological **anthropology**, or **molecular anthropology**, as well as a reference for **anthropologists** and anyone else interested in the genetic history of humans.

Chapter 14 - - atDNA results

Summary of the Autosomal testing

    T055537 James Lee Rader

    T579239 Earl Rader

    T829743 Harry Rader

    T918718 Alton Rader

How close are they related

The first test was processed by FTDNA with their relative finder. This test is like the one that's offered by ancestry.com and called their everything test. This test is an autosomal test. It looks at samples of your DNA from all chromosomes except for the Y chromosome. Our four candidates Alton Earl Jim and Harry are fourth cousins twice removed. The relationship between Harry and Jim are second cousin once removed, but they also have fourth cousin removed relationships on two other lines. We know these relationships through standard genealogical research which is well-documented.

Using the GEDmatch 3D Chromosome Browser, we get a nice chart showing the relationship between our four candidates. Earl and James share no DNA the rest of the relationships include one segment of the DNA. Harry and Jim share five segments because of their closer relationship.

Segments in common:

Kit	Name	T579239	T055537	T829743	T918718	Tot. Segments	Largest cM
T579239	Earl Rader	-	0	2	2	4	31.8
T055537	James Lee Rader	0	-	5	1	6	22.1
T829743	Harry Rader	2	5	-	1	8	22.1
T918718	Alton Rader	2	1	1	-	4	31.8

Autosomal DNA shared between pairs of relatives at the fourth cousin level normally share 13 cM but the fourth cousins once removed would only share 6 cM. So in our circumstance with the twice removed we would expect 3 cM. Using this table you can see the share five segments.

The expected level between Harry and Jim are that of second cousins once removed which on average 106 centiMorgans

Total Shared cM (Chr 1-22):

Kit	Name	T579239	T055537	T829743	T918718
T579239	Earl Rader	-	None	19.7	48.3
T055537	James Lee Rader	None	-	78.3	5.1
T829743	Harry Rader	19.7	78.3	-	11.4
T918718	Alton Rader	48.3	5.1	11.4	-

Summary by Chromosome:

hr	Tot. Matching Segments for all individuals.	B36 Graphic Posn Range		Largest segment
		From	To	
	0	72017	247185615	None
	0	8674	242697433	None
	1	36495	199322659	20.2
	1	49009	191200760	31.8
	0	78452	180630744	None
	0	100815	170761395	None
	0	139250	158812247	None
	2	154984	146264218	11.4
	2	36587	140208462	16.5
0	0	84172	135358259	None
1	0	188510	134445626	None
2	0	61880	132288869	None
3	0	17956717	114121631	None
4	2	18325726	106358708	21.6
5	1	18294933	100278685	9.2
6	0	28165	88690776	None
7	1	8547	78653169	5.1

8	0	3034	76116152	None
9	0	211912	63788972	None
0	0	9098	62382907	None
1	0	9849404	46924583	None
2	1	14494244	49558258	22.1
3	0	2321	154889941	None

Segment Details:

Kit1		Kit2		Chr	Build 36		cM
Kit Nbr.	Name	Kit Nbr.	Name		From	To	
T055537	James Lee Rader	T829743	Harry Rader	3	10215159	25460760	20.2
T579239	Earl Rader	T918718	Alton Rader	4	121202707	157186835	31.8
T055537	James Lee Rader	T829743	Harry Rader	8	140535029	143182478	5.2
T829743	Harry Rader	T918718	Alton Rader	8	135910061	141152797	11.4
T579239	Earl Rader	T918718	Alton Rader	9	129350655	136524980	16.5
T579239	Earl Rader	T829743	Harry Rader	9	136263069	140147760	13.6
T055537	James Lee Rader	T829743	Harry Rader	14	88480306	97899028	21.6
T579239	Earl Rader	T829743	Harry Rader	14	30421536	32553908	6.1
T055537	James Lee Rader	T829743	Harry Rader	15	88047217	91532308	9.2
T055537	James Lee Rader	T918718	Alton Rader	17	27056533	29380126	5.1
T055537	James Lee Rader	T829743	Harry	22	43130670	49528625	22.1

cM color coding	< 3 cM		Rader				
		3 - 5 cM	5 - 10 cM	10 - 20 cM	20 - 50 cM	50 - 100 cM	Over 100 cM

Triangulation Groups BETA Triangulation Groups – Expanded 600 kits   7 cm   all close relatives

Alton

FTDNA Chromosome browser

Alton Rader

Name	Match Date	Relationship Range	Shared Centimorgans	Longest Bloc
Earl Francis Rader Managed by James Lee Rader	12/28/2016	2nd Cousin - 4th Cousin	85	32
Harry Wayne Rader Managed by James Lee Rader	12/28/2016	5th Cousin - Remote Cousin	38	10

8

GEDMatch tools Alton to harry chrom 8

TG	Ref Kit	Name	Chr	Start	End	cM	SNPs
D47	T918718	T579239	Earl Rader	4	121196419	157197078	16.5 13865
D47	T918718	T579239	Earl Rader	9	129350356	136534626	16.5 3186

GEDMatch tools Comparing Kit T918718 (Alton Rader) and T579239 (Earl Rader)

Chr	Start Location	End Location	Centimorgans (cM)	SNPs
4	121,202,707	157,186,835	31.8	6,500
9	129,350,655	136,524,980	16.5	2,052

Earl

FTDNA Chromosome browser

By James Lee Rader

Earl is Alton's 4th cousin

  jim's   4th cousin – 2 R

  Harry's   4th cousin – 1 R

FTDNA 12/28/2016 Earl Rader is 2nd Cousin - 4th Cousin to Alton Rader 85 shared centimorgans, longest block 32

GEDMatch tools

TG	Ref	Kit	Name	Chr	Start	End	cM	SNPs
T579239	T829743		Harry Rader	9	136257591	140123767	13.6	1761

Name	Match Date	Relationship Range	Shared Centimorgans	Longest Block
Alton Rader Managed by James Lee Rader	12/28/2016	2nd Cousin - 4th Cousin	85	32

✉ ✎ ⬢

FTDNA Chromosome browser

How to do a Y-DNA study

NAME	MATCHNAME	CHROM	START LOCAT	END LOCATI	CENTIM	MATCHI
Earl Francis Rader	Alton Rader	1	60654943	62268948	2.87	500
Earl Francis Rader	Alton Rader	1	96069014	97915854	1.2	800
Earl Francis Rader	Alton Rader	2	33843174	35429477	1.91	500
Earl Francis Rader	Alton Rader	2	200674190	203562831	1.66	700
Earl Francis Rader	Alton Rader	3	130227425	132832117	2.09	500
Earl Francis Rader	Alton Rader	4	121698489	157150421	31.84	6791
Earl Francis Rader	Alton Rader	6	17535112	19579659	2.87	500
Earl Francis Rader	Alton Rader	6	29522265	32186335	1.67	3800
Earl Francis Rader	Alton Rader	6	32306027	33472061	2.6	1600
Earl Francis Rader	Alton Rader	7	86689033	88602336	2.6	700
Earl Francis Rader	Alton Rader	7	97170717	99702077	2.02	600
Earl Francis Rader	Alton Rader	8	49792703	53292222	2.19	600
Earl Francis Rader	Alton Rader	8	59884959	62623810	1.17	600
Earl Francis Rader	Alton Rader	9	129664544	136506361	7.7	2082
Earl Francis Rader	Alton Rader	11	48014889	58358885	3.36	1100
Earl Francis Rader	Alton Rader	11	87959264	90746870	1.39	500
Earl Francis Rader	Alton Rader	12	32965070	38203663	1.27	500
Earl Francis Rader	Alton Rader	12	59544596	61495890	2.74	500
Earl Francis Rader	Alton Rader	12	83512247	89243803	5.23	900
Earl Francis Rader	Alton Rader	15	45539006	47789753	1.85	500
Earl Francis Rader	Alton Rader	16	65173037	69077284	4.29	700

Total 84.52

GEDMatch tools

Comparing Kit T579239 (Earl Rader) and T829743 (Harry Rader)

Chr	Start Location	End Location	Centimorgans (cM)	SNPs
9	136,263,069	140,145,149	13.6	1,182

Harry Rader

Name	Match Date	Relationship Range	Shared Centimorgans	Longest Block
Brenda L Rader	11/15/2016	Parent/Child	3,384	267
Alton Rader Managed by James Lee Rader	12/28/2016	5th Cousin - Remote Cousin	38	10
James Lee Rader	11/15/2016	2nd Cousin - 4th Cousin	88	21

FTDNA Chromosome browser

NAME	MATCHNAME	CHROMC	START LOCA	END LOCATI	CENTIMC	MATCHII
Harry Wayne Rader	Alton Rader Mana	1	55430540	56984564	1.83	500
Harry Wayne Rader	Alton Rader Mana	2	102654532	105248493	2.67	500
Harry Wayne Rader	Alton Rader Mana	4	98167728	100861136	1.52	600
Harry Wayne Rader	Alton Rader Mana	5	128068207	132242229	2.6	800
Harry Wayne Rader	Alton Rader Mana	7	86689033	87899607	1.33	500
Harry Wayne Rader	Alton Rader Mana	8	65670476	68010769	3.01	500
Harry Wayne Rader	Alton Rader Mana	8	136075812	140875208	9.95	1400
Harry Wayne Rader	Alton Rader Mana	11	46718718	58858258	3.87	1400
Harry Wayne Rader	Alton Rader Mana	12	20668638	22538447	4.65	900
Harry Wayne Rader	Alton Rader Mana	17	47625736	50121209	3.57	500
Harry Wayne Rader	Alton Rader Mana	20	24539028	30554896	2.59	500
					37.59	8100
Harry Wayne Rader	James Lee Rader	2	38003856	40202666	1.86	600
Harry Wayne Rader	James Lee Rader	3	9519554	25394306	21.33	4800
Harry Wayne Rader	James Lee Rader	3	49899944	53851258	2.02	700
Harry Wayne Rader	James Lee Rader	6	137601988	139406923	2.48	554
Harry Wayne Rader	James Lee Rader	7	86131629	87246282	1.47	500
Harry Wayne Rader	James Lee Rader	8	50532761	54202788	2.5	700
Harry Wayne Rader	James Lee Rader	8	140646001	143076563	3.64	700
Harry Wayne Rader	James Lee Rader	9	111725121	113086729	2.88	500
Harry Wayne Rader	James Lee Rader	10	73429800	76701795	2.05	500
Harry Wayne Rader	James Lee Rader	10	104404211	106318781	1.86	500
Harry Wayne Rader	James Lee Rader	12	20668638	21368250	2.53	500
Harry Wayne Rader	James Lee Rader	12	32965070	39002106	1.8	700
Harry Wayne Rader	James Lee Rader	12	69386568	71634812	2.54	500
Harry Wayne Rader	James Lee Rader	14	88516549	97762749	19.37	3171
Harry Wayne Rader	James Lee Rader	15	88174443	91527938	10.03	1233
Harry Wayne Rader	James Lee Rader	18	23287051	25213638	1.71	500
Harry Wayne Rader	James Lee Rader	22	43142315	45772802	7.55	1005
					87.62	17663

GEDMatch tools

TG	Ref	Kit	Name	Email	Chr	Start	End	cM	SNPs
C07	T829743	T055537	James Lee Rader		3	9182331	25555780	23.1	8919
C07	T829743	T055537	James Lee Rader		14	88176648	98064968	23.1	5652
C07	T829743	T055537	James Lee Rader		15	87808594	91637377	23.1	2355
C07	T829743	T055537	James Lee Rader		22	42945646	49542594	23.1	4302

Chr	Start Location	End Location	Centimorgans (cM)		SNPs
3	9,325,167	25,460,760	21.4	4,680	
14	88,480,306	97,899,028	21.6	3,106	
15	88,047,217	91,532,308	9.1	1,234	
22	43,130,670	49,528,625	22.1	2,528	

Jim

Harry Wayne Rader Managed by James Lee Rader	11/15/2016	2nd Cousin - 4th Cousin	88	21

FTDNA Chromosome browser

How to do a Y-DNA study

NAME	MATCHNAME	CHROM	START LOCATIO	END LOCATIO	CENTIMO	MATCHI
James Lee Rader	Harry Wayne Rader	2	38003856	40202666	1.86	600
James Lee Rader	Harry Wayne Rader	3	9519554	25394306	21.33	4800
James Lee Rader	Harry Wayne Rader	3	49899944	53851258	2.02	700
James Lee Rader	Harry Wayne Rader	6	137601988	139406923	2.48	554
James Lee Rader	Harry Wayne Rader	7	86131629	87246282	1.47	500
James Lee Rader	Harry Wayne Rader	8	50532761	54202788	2.5	700
James Lee Rader	Harry Wayne Rader	8	140646001	143076563	3.64	700
James Lee Rader	Harry Wayne Rader	9	111725121	113086729	2.88	500
James Lee Rader	Harry Wayne Rader	10	73429800	76701795	2.05	500
James Lee Rader	Harry Wayne Rader	10	104404211	106318781	1.86	500
James Lee Rader	Harry Wayne Rader	12	20668638	21368250	2.53	500
James Lee Rader	Harry Wayne Rader	12	32965070	39002106	1.8	700
James Lee Rader	Harry Wayne Rader	12	69386568	71634812	2.54	500
James Lee Rader	Harry Wayne Rader	14	88516549	97762749	19.37	3171
James Lee Rader	Harry Wayne Rader	15	88174443	91527938	10.03	1233
James Lee Rader	Harry Wayne Rader	18	23287051	25213638	1.71	500
James Lee Rader	Harry Wayne Rader	22	43142315	45772802	7.55	1005

TG	Ref	Kit	Name I	Chr	Start	End	cM	SNPs
C07	T055537	T829743	Harry Rader	3	9320218.00	25465973.00	22.1	8719
C07	T055537	T829743	Harry Rader	14	88480306.00	97902070.00	22.1	5452
C07	T055537	T829743	Harry Rader	15	88047217.00	91534413.00	22.1	2155
C07	T055537	T829743	Harry Rader	22	43130670.00	49542594.00	22.1	4202

GEDMatch tools

GEDmatch.Com Autosomal Comparison - V2.1.1(c)

Comparing Kit T055537 (James Lee Rader) and T829743 (Harry Rader)

Chr	Start Location	End Location	Centimorgans (cM)	SNPs
3	9,325,167	25,460,760	21.4	4,680
14	88,480,306	97,899,028	21.6	3,106
15	88,047,217	91,532,308	9.1	1,234
22	43,130,670	49,528,625	22.1	2,528

Appendix

Compared all STRS table

	Jim 3497	Harry 7939	Earl 8017	Alton 8525	
ATA71D03	n/a	n/a	n/a	n/a	###
CDY.1	34	n/a	n/a	n/a	###
CDY.2	37	n/a	n/a	n/a	###
DXYS156	12	n/a	n/a	12	###
DYF371.1	10	10	10	10	10
DYF371.2	12	12	12	12	12
DYF371.3	13	13	13	13	13
DYF371.4	14	14	14	14	14
DYF380.1	10	10	10	10	10
DYF380.2	10	10	10	10	10
DYF381.1	8	8	8	8	8
DYF381.2	8	8	8	8	8
DYF382	n/a	n/a	n/a	n/a	###
DYF383.1	9	n/a	9 ?	9	###
DYF383.2	9	n/a	9 ?	9	###
DYF384.1	7	7	7	7	7
DYF384.2	8	8	8	8	8
DYF385.1	10	n/a	10	10	###
DYF385.2	11	n/a	11	11	###
DYF386.1	11	11 ?	14 ?	11 ?	12
DYF386.2	14	11 ?	14 ?	11 ?	13
DYF386.3	14	14 ?	14 ?	14 ?	14
DYF386.4	14	14 ?	14 ?	14 ?	14
DYF387.1	31	31	31	31	31
DYF387.2	31	31	31	31	31
DYF389	11	11	11	11	11
DYF391.1	9	9	9	9	9
DYF391.2	9	9	9	9	9
DYF392	8	n/a	8	8	###
DYF393	27	27	27	27	27
DYF394	8	n/a	n/a	n/a	###
DYF395.1	16	16	16	16	16
DYF395.2	16	16	16	16	16
DYF396.1	8	8	8	8	8
DYF396.2	8	8	8	8	8

	Jim	Harry	Earl	Alton	
	3497	7939	8017	8525	
DYF398.1	8	8	8	8	8
DYF398.2	17	17	17	17	17
DYF399.1	23 ?	24.1 ?	23 ?	23 ?	23
DYF399.2	24.1 ?	25 ?	24.1 ?	24.1 ?	24
DYF399.3	25 ?	25 ?	26 ?	25 ?	25
DYF400.1	22	n/a	22 ?	22	###
DYF400.2	22	n/a	22 ?	22	###
DYF401.1	14	14	14	14	14
DYF401.2	16	16	16	16	16
DYF403.1	26	n/a	25 ?	26 ?	###
DYF403.2	28	n/a	26 ?	28 ?	###
DYF404.1	14	13 ?	14	14	14
DYF404.2	16	16 ?	16	16	16
DYF405.1	6	6	6	6	6
DYF405.2	11	11	11	11	11
DYF406	10	10	10	10	10
DYF407.1	12	12	12	12	12
DYF407.2	12	12	12	12	12
DYF408.1	8	8	8	8	8
DYF408.2	14	14	14	14	14
DYF409.1	11	11	11	11	11
DYF409.2	13	13	13	13	13
DYF410.1	10	n/a	10	10	###
DYF410.2	11	n/a	11	11	###
DYF411.1	11	n/a	11 ?	11	###
DYF411.2	11	n/a	12 ?	11	###
DYF412.1	11	11	11	11	11
DYF412.2	12	12	12	12	12
DYR1	17	n/a	17	17 ?	###
DYR10	10	10	10	10	10
DYR100	10	n/a	10	10	###
DYR101	10	10	10	n/a	###
DYR102	7	n/a	n/a	n/a	###
DYR103	n/a	n/a	n/a	n/a	###
DYR104	9	n/a	9	9	###

	Jim 3497	Harry 7939	Earl 8017	Alton 8525	
DYR105	5	n/a	5	n/a	###
DYR106	8	8	8	8	8
DYR107	n/a	n/a	n/a	n/a	###
DYR108	9	9	9	9	9
DYR110	5	5	5	5	5
DYR111	12	12	12	12	12
DYR112	13	n/a	13	13	###
DYR113	6	6	6	6	6
DYR114	n/a	n/a	n/a	n/a	###
DYR115	29	n/a	n/a	n/a	###
DYR116	4	4	4	4	4
DYR117	7	7	7	7	7
DYR118	10	n/a	10	10	###
DYR119	13	13	13	13	13
DYR12	10	10	10	10	10
DYR120	6	n/a	6	6	###
DYR121.1	14	14 ?	14 ?	14	14
DYR121.2	15	15 ?	15 ?	14	15
DYR122.1	7	7	7	7	7
DYR122.2	7	7	7	7	7
DYR123	16	n/a	16	16	###
DYR124.1	9	n/a	9 ?	9	###
DYR124.2	10	n/a	9 ?	9	###
DYR124.3	12	n/a	9 ?	12	###
DYR125.1	23	23	23	23	23
DYR125.2	26	26	26	26	26
DYR126	12	12	12	12	12
DYR127	n/a	n/a	n/a	n/a	###
DYR128.1	10	10	10	10	10
DYR128.2	10	10	10	10	10
DYR13	12	12	12	12	12
DYR130	13	13	13	12 ?	13
DYR131	7	7	7	7	7
DYR132.1	n/a	n/a	12 ?	12	###
DYR132.2	n/a	n/a	12.3 ?	12.3	###

	Jim	Harry	Earl	Alton	
	3497	7939	8017	8525	
DYR135	10	10	10	10	10
DYR136	6	6	6	6	6
DYR137	4	4	4	4	4
DYR138	20	20	21	20	20
DYR139	16	n/a	17	17	###
DYR14	11	11	11	11	11
DYR143	11	11	11	11	11
DYR144	9	9	9	9	9
DYR146	9	9	9	9	9
DYR15	12	12	12	12	12
DYR150	13	13	13	13	13
DYR152	10	n/a	n/a	10	###
DYR154	12	12	12	12 ?	12
DYR156	21	21	21	21	21
DYR157	13	13	13	13	13
DYR158	11	n/a	11	11	###
DYR159	16	16	16	16	16
DYR160	13	n/a	13	13	###
DYR161	15	n/a	n/a	15	###
DYR162	16	n/a	16	16	###
DYR163	31	31	31	31	31
DYR164	25	n/a	25	25	###
DYR165	n/a	n/a	45	44	###
DYR166	n/a	11	11	11	###
DYR167	n/a	11.t	11.t	n/a	###
DYR168	n/a	6	6	6	###
DYR169	n/a	8	8	n/a	###
DYR17.1	12	12	12	12	12
DYR17.2	13	13	13	13	13
DYR17.3	14	14	14	14	14
DYR170	n/a	n/a	36 ?	36	###
DYR171	n/a	n/a	43.g	43.g ?	###
DYR172	n/a	17	17	17	###
DYR173	n/a	n/a	n/a	27	###
DYR174	n/a	13	13	13	###

	Jim	Harry	Earl	Alton	
	3497	7939	8017	8525	
DYR175	n/a	n/a	15	15	###
DYR18.1	10	n/a	10	n/a	###
DYR18.2	14	n/a	14	n/a	###
DYR19	10	n/a	10	10	###
DYR2	12	12	12	12	12
DYR20	10.t	10.t ?	10.t	10.t	###
DYR23	12	n/a	n/a	12	###
DYR26	12	12	12	12	12
DYR27	n/a	n/a	n/a	n/a	###
DYR28	11	11	11	11	11
DYR29	n/a	n/a	n/a	n/a	###
DYR3	10	10	10	10	10
DYR30	15	n/a	n/a	15	###
DYR31	7	7	7	7	7
DYR32	n/a	n/a	n/a	18	###
DYR33	15	15	15	15	15
DYR35.1	9	9 ?	9	9	9
DYR35.2	9	9 ?	9	9	9
DYR36.1	4	4	4	4	4
DYR36.2	7	7	7	7	7
DYR38.1	12	n/a	12	12	###
DYR38.2	13	n/a	13	13	###
DYR39	11	11	11	11	11
DYR40	9	9	8	9	8.8
DYR41	n/a	n/a	n/a	11	###
DYR43	9	9	9	9	9
DYR44	11	n/a	12	11	###
DYR45.1	10 ?	10 ?	10	10	10
DYR45.2	11 ?	10 ?	11	11	11
DYR45.3	13.g ?	11 ?	13.g	13.g	###
DYR46	9	9	9	n/a	###
DYR47	4	4	4	4	4
DYR48	11	11	n/a	11	###
DYR49	n/a	n/a	n/a	n/a	###
DYR5	8	n/a	n/a	n/a	###

	Jim	Harry	Earl	Alton	
	3497	7939	8017	8525	
DYR51	7	7	7	7	7
DYR52	9	9	9	n/a	###
DYR54	10	10	10	10	10
DYR55	12	n/a	12	12 ?	###
DYR56	16	16	16	16	16
DYR57	n/a	n/a	11	11	###
DYR58.1	9	n/a	9	9	###
DYR58.2	12	n/a	12	12	###
DYR59	n/a	n/a	n/a	n/a	###
DYR6	15	15	15	15	15
DYR60	18	18	18	n/a	###
DYR61	n/a	n/a	n/a	n/a	###
DYR62	n/a	n/a	n/a	n/a	###
DYR63.1	12	12 ?	12	12	12
DYR63.2	13	13 ?	13	13	13
DYR64.1	n/a	n/a	n/a	n/a	###
DYR64.2	n/a	n/a	n/a	n/a	###
DYR65	8	n/a	8	n/a	###
DYR66.1	10	10	10	10	10
DYR66.2	10	10	10	10	10
DYR67.1	9	9	9	9	9
DYR67.2	9	9	9	9	9
DYR67.3	11	11	11	11	11
DYR67.4	11	11	11	11	11
DYR68.1	9	9 ?	9	9	9
DYR68.2	9	9 ?	9	9	9
DYR68.3	9	9 ?	13	13	11
DYR68.4	13	13 ?	13	13	13
DYR69	n/a	n/a	n/a	n/a	###
DYR7	14	14	14	14	14
DYR70	10	n/a	10	10	###
DYR71	7	7	7	7	7
DYR73	10	10	10	10	10
DYR74	11	11	11	11	11
DYR75	13	n/a	13 ?	13	###

	Jim 3497	Harry 7939	Earl 8017	Alton 8525	
DYR76	13	13	13	13	13
DYR77	11	11	11	11	11
DYR78	n/a	n/a	n/a	n/a	###
DYR79	16	18	16	16	17
DYR8	11	n/a	11	11	###
DYR80	6	6	6	6	6
DYR81	34	34	34	34	34
DYR82	10	10	10	10	10
DYR83	12	12 ?	12	12	12
DYR84	15	15	15	14	15
DYR85	8	8	8	8	8
DYR87	7	n/a	7	7	###
DYR88.1	17	n/a	16 ?	17 ?	###
DYR88.2	19	n/a	18 ?	18 ?	###
DYR89	11	11	11	11	11
DYR9.1	13	n/a	13 ?	13 ?	###
DYR9.2	14	n/a	14 ?	14 ?	###
DYR90	13	13	14	13	13
DYR91	6	6	6	6	6
DYR92	12	12	12	12	12
DYR93	13	13	13	13	13
DYR94	10	n/a	10 ?	10	###
DYR95	7	7	7	7	7
DYR96	7	n/a	7	n/a	###
DYR97	13	n/a	12	12	###
DYR99	7	n/a	7	7	###
DYS19	14	14	14	14	14
DYS385.1	11	n/a	11	11	###
DYS385.2	14	n/a	14	14	###
DYS388	12	12	12	12	12
DYS389I	13	13	13	13	13
DYS389II	29	29	29	29	29
DYS390	24	24	24	24	24
DYS391	11	11	11	11	11
DYS392	13	n/a	13	13	###

How to do a Y-DNA study

	Jim	Harry	Earl	Alton	
	3497	7939	8017	8525	
DYS393	13	13	13	13	13
DYS413.1	23	n/a	n/a	n/a	###
DYS413.2	23	n/a	n/a	n/a	###
DYS425	12	12	12	12	12
DYS426	12	12	12	12	12
DYS434	9	9	9	9	9
DYS435	11	11	11	11	11
DYS436	12	12	12	12	12
DYS437	15	15	15	15	15
DYS438	12	12	12	12	12
DYS439	12	12	12	12	12
DYS441	13	13	13	13	13
DYS442	13	13	13	13	13
DYS443	13	13	13	13	13
DYS444	11	11	11	11	11
DYS445	12	12	12	12 ?	12
DYS446	14	14	14	14	14
DYS447	25	n/a	n/a	n/a	###
DYS448	19	19	19	19	19
DYS449	n/a	n/a	n/a	30	###
DYS450	8	n/a	8	8	###
DYS452	30	n/a	n/a	n/a	###
DYS453	11	n/a	11	11	###
DYS454	11	n/a	11	11	###
DYS455	11	n/a	11	n/a	###
DYS456	16	16	16	16	16
DYS458	17	17	17	17	17
DYS459.1	9	9	9	9	9
DYS459.2	9	9	9	9	9
DYS460	11	11	11	11	11
DYS461	12	12	12	12	12
DYS462	11	11	11	n/a	###
DYS463	n/a	n/a	n/a	n/a	###
DYS464.1	15	15 ?	15	15	15
DYS464.2	15	15 ?	15	15	15

	Jim	Harry	Earl	Alton	
	3497	7939	8017	8525	
DYS464.3	17	17 ?	17	17	17
DYS464.4	18	18 ?	18	18	18
DYS466	7	7	7	7	7
DYS467	12	12	12	12	12
DYS468	17	17	17	17	17
DYS469	16	16	16 ?	16	16
DYS470	11	11	11	11	11
DYS471	28	28	28	28	28
DYS472	8	8	8	8	8
DYS473	n/a	n/a	n/a	n/a	###
DYS474	8	8	8	8	8
DYS475	8	8	8	8	8
DYS476	11	11	11	11	11
DYS477	8	8	8	8	8
DYS478	n/a	n/a	n/a	n/a	###
DYS480	8	8	8	8	8
DYS481	22	22	22	22	22
DYS484	13	n/a	13	13	###
DYS485	16	16	16	16	16
DYS487	n/a	13	n/a	n/a	###
DYS488	13	n/a	13	13	###
DYS489	11	11	11	11	11
DYS490	n/a	n/a	n/a	n/a	###
DYS491	13	14	n/a	13	###
DYS492	12	n/a	12	12	###
DYS493	10	10	10	10	10
DYS494	9	9	9	9	9
DYS495	16	n/a	16	16	###
DYS496	n/a	n/a	n/a	n/a	###
DYS497	14	n/a	n/a	14	###
DYS499	8	8	8	8	8
DYS500	n/a	n/a	14 ?	13	###
DYS501	7	n/a	7	7	###
DYS502	13	13	13	13	13
DYS504	17	17	17	17	17

How to do a Y-DNA study

	Jim 3497	Harry 7939	Earl 8017	Alton 8525	
DYS505	12	n/a	n/a	12	###
DYS506	13	13	n/a	13	###
DYS507	10	10	10	10	10
DYS508	11	11	11	11	11
DYS509	n/a	n/a	n/a	n/a	###
DYS510	17	17	17	17	17
DYS511	11	11	11	11	11
DYS512	11	11	11	11	11
DYS513	11	11	11	11	11
DYS514	21	21	21	21	21
DYS516	12	12	12	12	12
DYS517	15	n/a	15	15	###
DYS518	33	n/a	33	33	###
DYS520	20	20	20	20	20
DYS521	9	9	9	9	9
DYS522	11	11	11	11	11
DYS523	15	15	15	15	15
DYS525	10	10	10	10	10
DYS526A	14	n/a	14	n/a	###
DYS526B	36	n/a	36	36 ?	###
DYS527.1	33	n/a	n/a	33	###
DYS527.2	35	n/a	n/a	35	###
DYS528.1	19	18 ?	18	18	18
DYS528.2	19	19 ?	19	19	19
DYS530	9	9	9	9	9
DYS531	11	11	11	11	11
DYS532	13	n/a	13	13	###
DYS533	12	12	12	12	12
DYS534	16	16	16	16	16
DYS536	10	n/a	10	n/a	###
DYS537	10	10	10	10	10
DYS538	10	10	10	10	10
DYS539	10	10	10	10	10
DYS540	13	n/a	13	n/a	###
DYS541	12	12	12	12	12

By James Lee Rader

	Jim	Harry	Earl	Alton	
	3497	7939	8017	8525	
DYS542	14	14	14	14	14
DYS543	17	17	17	17	17
DYS544	14	14	14	14	14
DYS545	10	10	10	10	10
DYS546	17	17	17	17	17
DYS547	42	n/a	42	42	###
DYS548	13	n/a	13	n/a	###
DYS549	12	12	12	12	12
DYS550	8	8	8	8	8
DYS551	12	12	12	12	12
DYS552	24	24	24	24	24
DYS554	9	n/a	9	9	###
DYS556	11	11	11	11	11
DYS557	16	16	16	16	16
DYS558	n/a	n/a	n/a	n/a	###
DYS559	8	n/a	8	8	###
DYS561	15	n/a	15	15	###
DYS562	20	20	20	20	20
DYS565	n/a	n/a	n/a	n/a	###
DYS567	11	n/a	11	11	###
DYS568	n/a	n/a	11	11	###
DYS569	11	11	11	11	11
DYS570	17	17	17	17	17
DYS571	10	n/a	10	10	###
DYS572	n/a	n/a	n/a	11	###
DYS573	10	10	10	10	10
DYS574	10	10	10	10	10
DYS575	10	10	10 ?	10	10
DYS576	16	16	16	16	16
DYS577	9	9	9	9	9
DYS578	9	9	9	9	9
DYS579	9	9	9	9	9
DYS580	9	9	9	9	9
DYS581	8	8	8	8	8
DYS582	8	n/a	8	8	###

How to do a Y-DNA study

	Jim	Harry	Earl	Alton	
	3497	7939	8017	8525	
DYS583	8	8	8	8	8
DYS584	8	8	8	8	8
DYS585	10	10	10	9	9.8
DYS587	19	19	19	19	19
DYS588	12	12	12	12	12
DYS589	12	n/a	12	12	###
DYS590	8	8	8	8	8
DYS592	11	n/a	11	11	###
DYS593	15	n/a	15	15	###
DYS594	10	10	10	10	10
DYS595	8	n/a	n/a	8	###
DYS596	10	n/a	10	10	###
DYS598	7	7	7	7	7
DYS599	22	n/a	22	22	###
DYS600	11	11	11	11	11
DYS607	15	15	15	15	15
DYS608	9	9	9	9	9
DYS609	n/a	n/a	n/a	n/a	###
DYS611	17	n/a	18 ?	17	###
DYS612	30	30	32	32	31
DYS613	8	8	8	8	8
DYS614	18	18	n/a	18	###
DYS615	8	8	8	8	8
DYS616	14	14	14	14	14
DYS617	n/a	n/a	n/a	n/a	###
DYS618	12	12	12	12	12
DYS619	n/a	n/a	n/a	n/a	###
DYS620	8	8	8	8	8
DYS621	8	n/a	8	n/a	###
DYS622	20	20	20	20	20
DYS623	10	10	10	10	10
DYS624	9	9	9	9	9
DYS625	33	33	33	33	33
DYS626	26	n/a	25	27 ?	###
DYS627	28	28	28	28	28

By James Lee Rader

	Jim	Harry	Earl	Alton	
	3497	7939	8017	8525	
DYS629	n/a	n/a	n/a	n/a	###
DYS630	22	22	22	22	22
DYS631	10	n/a	10	10	###
DYS632	9	9	9	9	9
DYS633	n/a	n/a	n/a	n/a	###
DYS634	8	8	8	8	8
DYS635	23	23	23	23	23
DYS636	12	12	13	12	12
DYS637	11	11	11	11	11
DYS638	11	11	11	11	11
DYS639	10	n/a	10	10	###
DYS640	11	11	11	11	11
DYS641	10	n/a	10	10	###
DYS642	7	n/a	7	7	###
DYS643	10	10	10	10	10
DYS644	16	16	16	16	16
DYS645	8	8	8	8	8
DYS649	8	8	8	8	8
DYS650	20	20	20	20	20
DYS651	n/a	n/a	n/a	n/a	###
DYS655	n/a	n/a	n/a	n/a	###
DYS656	37	n/a	n/a	36	###
DYS662	n/a	n/a	8	8	###
DYS664	50	n/a	50 ?	50	###
DYS666	7	7	7	7	7
DYS667	7	7	7	7	7
DYS668	5	5	5	5	5
DYS672	9	9	9	9	9
DYS673	11	n/a	n/a	n/a	###
DYS675	6	6	6	6	6
DYS676	11	11	11	11	11
DYS677	11	11	11	11	11
DYS678	11	11	n/a	11	###
DYS679	12	12	12	12	12
DYS681	5	n/a	5	n/a	###

	Jim 3497	Harry 7939	Earl 8017	Alton 8525	
DYS683	10	10	10	10	10
DYS684	n/a	n/a	57 ?	n/a	###
DYS685	42	n/a	n/a	43	###
DYS686	21	n/a	21	n/a	###
DYS687	34	n/a	n/a	34	###
DYS688	80	n/a	n/a	79 ?	###
DYS692	7	7	7	7	7
DYS694	8	n/a	8	8	###
DYS695	32	32 ?	32 ?	32	32
DYS696	7	7	7	7	7
DYS701	5	5	5	5	5
DYS702	n/a	n/a	30	30.t	###
DYS703	10	n/a	10	10	###
DYS705	5	5	5	5	5
DYS706	14	14	14	14	14
DYS707	n/a	23	23	n/a	###
DYS708	28	28	28	28	28
DYS709	20	20	20	20	20
DYS710	35	n/a	n/a	35.2	###
DYS711	66	n/a	n/a	65	###
DYS712	22	22	22	21	22
DYS713	45	n/a	n/a	46	###
DYS714	26	26	26	26	26
DYS715	24	24	24	24	24
DYS716	n/a	n/a	n/a	n/a	###
DYS717	19	19	19 ?	n/a	###
DYS718	15	n/a	15	15	###
DYS719	13	14	n/a	13	###
DYS720	31	n/a	31	31	###
DYS721	19	19	19	19	19
DYS722	22	n/a	22	22	###
DYS723	19	19	19	19	19
DYS725.1	n/a	n/a	30 ?	31 ?	###
DYS725.2	n/a	n/a	31 ?	31 ?	###
DYS725.3	n/a	n/a	31 ?	32 ?	###

	Jim	Harry	Earl	Alton	
	3497	7939	8017	8525	
DYS725.4	n/a	n/a	32 ?	32 ?	###
DYS726	12	12	12	12	12
G09411	11	11	11	n/a	###
L1313	3	3	3	3	3
L14	4	4	4	4	4
YCAII.1	19	n/a	n/a	n/a	###
YCAII.2	23	n/a	n/a	n/a	###
Y-GATA-A10	12	12	12	12	12
Y-GATA-H4	11	11	11	11	11
Y-GGAAT-1B07	10	10	10	10	10

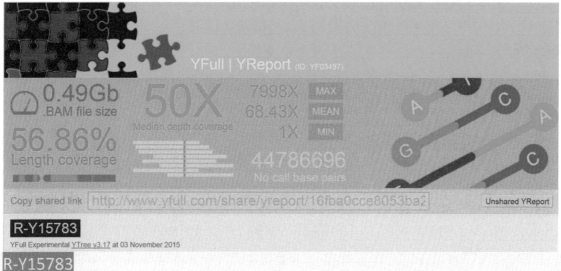

YFull | YReport (ID: YF03497)

0.49Gb .BAM file size

50X Median depth coverage

7998X MAX
68.43X MEAN
1X MIN

56.86% Length coverage

44786696 No call base pairs

Copy shared link http://www.yfull.com/share/yreport/16fba0cce8053ba2

Unshared YReport

R-Y15783

YFull Experimental YTree v3.17 at 03 November 2015

R-Y15783

YFull Experimental YTree v3.11 at 06 June 2015

ROOT (Y-Chromosome "Adam")

A1 Z8743 Y13319 L985 YP3855 YP3854 YP3857 YP3845 YP3843 YP3862 A4739 •
YP3847 A4826 •
YP3865 YP3846 Z8529 Y13320 L989 L1009 Y14726 V171 V238 L1053 YP3851 YP3849 V241
L986 Z8830 YP3848 YP3859 V250 YP3853L1002 M9072 YP3856 A4719 YP3866 P305 YP386
4 Z9582 V161 • V161.1 • V161.2 L1112 YP3863 A4733 Z9656 YP3867 A4698 •
YP3839 L1084 L1004 Z9116 V168 YP3840 YP3852 YP3850 YP3858 A4825 •
YP3861 Z9072 V174 YP3841 Z9045YP3868 L1005 YP3860 Z8834 YP3842 A4709 •
Y13318 Y1459 YP3844
A1b Y8296 Y9420 Y10869 Y8894 Y10883 Y8284 V221 Y10840 Y10877 Y10854 P108 Y82
92 Y10860 Y10844 Y8291 Y8283 Y8290 Y10863 Y8295 Y8297 Y8286 Y8278 Y8294 Y8293 Y1
0882 Y8302 Y8300 Y10870 Y10848 Y8279Y10865 Y8289 Y8288 Y8303 Y8301 Y8285 Y8280 Y
10864 Y8287 Y8282 Y10857 Y8281 Y8298 Y10850 Y10856 Y8299 Z11900 • Y8277
BT V2952 • M9137 M9159 • PF767 M9312 M9380 • PF1256 M9382 •
PF1257 Y10861 V2167 • M9026 M9178 M9251 • PF913 M9113 M9133 •
PF715 M8952 M9232 • PF880 M8961 • PF201 M9202 V3063 • M9145 •
PF733 M8972 M9221 M299M9356 M9070 Z11946 • M8967 M8968 • PF207 Z17342 •
Y9452 V2804 • M9124 • PF701 M9157 • PF766 M9081 Y8488 Y10843 M9378 M9379 •
PF1253 V2561 • M9110 • PF684 M9054 M9036 •
PF308 M9349 M9361 M9396 M9214 M9287M9365 • PF1218 M9226 •
PF869 M9057 M9389 M9338 • PF1064 M9359 M9046 • PF324 M9322 • PF1049 L970 •
PF1065 M9027 V2319 • M9031 M9203 • PF837 M9238 M9346 L957 M9123 M9240 •

PF896 M9305 • PF1022 Z17365 • Y10859M9372 V4201 • M9297 • PF1003 M9103 •

PF679 Y10868 M9292 • PF995 M9249 M8958 • PF196 M9321 •

PF1045 M9375 M9406 M9278 • PF969 M9188 Z17362 • Y10885 V1395.1 • M8997 •

PF260 M9099 • PF674 M9021 • PF288 M9104 • PF680M9002 • PF267 V2579 •

M9114 Z17336 • Y9392 M9177 Z17354 • Y10845 V1158 • M8954 M9155 •

PF762 M9174 M9329 M8980 • PF229 M9311 • PF1030 M9204 M9231 • PF876 A4807 •

Y1547 2 M9197 M9254 M9230 • PF870 M9102 M9135Z17337 • Y9417 M9394 •

PF1271 M9246 V2209 • M9028 • PF298 M9083 M9196 M9098 M9213 M9041 •

PF319 M9326 M9366 PF1407 • V21 • M8969 V2507 • M9038 • PF313 V4025 •

M9291 M9069 • PF635 M9187 M9411 • PF1315 L1060 • PF1021 M9412 M9200 •

PF835 M9223 • PF865 M9227 V1813 • M9019 • PF286 M9237 • PF890 M9172 Z12093 •

V4213 • M9298 V3107 • M9148 • PF744 M9220 M8951 V3636 • M9269 SRY10831 •

PF6234 • PAGE65.1 • SRY1532.1 • SRY10831.1V2437 • M9034 V3226 •

M9152 M9138 Z17357 • Y10871 V1347 • M8994 M9357 • PF1209 P97 L1071 •

M8945 M9141 Y9451 M9136 • PF724 V2821 • M9126 • PF703 M9191 M9370 M9318 •

PF1039 M9327 Y10875 M8959 • PF198 M9105 M9068M8983 • PF230 M9179 M9352 •

PF1100 Z17346 • Y10889 M9209 M9244 M9042 Z17386 • Y6870 PF1405 • V216 •

M8953 Z17340 • Y10842 Y10876 PAGES00024 • M9160 Y10853 M9334 M9343 •

PF1084 M9003 V64 • PF1412 Y15547 V2352 • L1062 • PF302 M8976 • PF215 M9266 •

PF946 V202 • PF1404 M8985 • PF232 Y9450 M9074 Z17348 •

Y10880 M8993 M9360 M9405 M9112 Z17344 • Y9394 M9146 M9225 • PF868 V1730 •

M9017 • PF282 M9193 L440 A4808 • Y1546 2 V29 • PF1408 V31 • L413 •

PF1409 M8971 Y9422 M9302 V2397 • M9032 • PF304 Y10888 M42 L438 M9393 M8955 •

PF12 M9189 M9425 M9130 • PF708 Y9131 Y8489 M9107 Y10841 L418 M9015 V2000 •

M9025 M8960 • PF200 M9075PAGES00026 • M9336 M9094 • PF671 M9219 •

CTS7503 M9115 • PF687 M8988 M9290 • PF989 M9176 M9192 M9169 V2318 •

M9030 V3304 • M9257 V3601 • M9267 • PF948 M9195 L971 M9011 M9284 M9331 •

PF1057 A5288 • M9420 M9217 • PF857 M9001 M9319 Y10884 M9216 V3998 • M9288 •

PF985 PF601 Z17356 • Y10852 M8999 M9286 Z17359 • Y10847 V3297 • M9255 •

PF925 M9304 V3037 • M9143 • PF732 M9228 M9039 V3002 •

M9140 M9280 M9353 L1061 • PF1101Y9419 M9020 • PF287 M9077 Y10839 M8979 •

PF226 V3916 • M9283 • PF973 Y11581 M9180 M91 M9315 • PF1033 Z17341 •

Y10879 M9367 Z17371 • Y8487 M9111 M9165 L977 M9100 M9121 M9199 •

PF834 Y10837 M9376 M9236 M9248Y10866 M9272 M9210 M9245 Z17349 •

Y10849 M9016 Z17372 • Y10862 M9348 • PF1093 PF1406 • V102 M9182 M9215 •

PF847 V59 • PF1411 M9089 • PF653 V187 • PF1403 Y10855 M9260 M9317 M9004 •

PF270 M9235 •

PF886 M9043M9056 M9125 Y10878 Y10851 M9344 M9368 M9066 M9325 •

PF1052 M9087 L1220 • M9212 M9234 • PF885 M9285 M9296 V1456 • M9000 V2656 • PAGES00081 • M9118 M9131 V3347 • M9261 • PF931 V3032 • M9142 • PF731 M9390 • PF1262 M9166 • PF785 M8956 • PF14 M9408 • PF1296 M9335 • PF1060 M9306 Z17339 • Y9393 M9323 • PF1050 M8977 M9006 M139 M9404 M9095 M9050 V4007 • M9289 • PF988 M94 • PF1081 Y10873 M8973 • PF211 V3795 • M9277M9303 M9301 • PF1015 M9175 M9258 V3357 • M9262 • PF932 M9417 Z17345 • Y10872 Y9418 M9128 M9224 L962 Y10838 M9347 V1506 • M9005 V3904 • M9282 M9080 M9253 • PF914 V1015 • M8947 M9156 • PF764 M8957 M9400 • PF1284 M9065 • PF351 V4130 • M9295 • PF1000 L604 • PF1243 Y10874 M9374 • Z4690 A5289 • M9421 M9076 V2760 • PF699 • M251 • M9122 M9151 M9300 M9139 M9270 • PF952 M9163 • PF777 M8970 • PF208 M9218 • PF860 M9362 M9239M9109 M9263 V3546 • M9265 M9293 • PF997 Z17343 • L978 • PF93 M9409 • CTS12197 • PF1314 M8986 M9045 M9086 • PF648 M9369 M9310 M9328 • PF1053 M9127 V2634 • M9117 M9399 • PF1283 M9373 M9198 M9242 • PF899 Z17355 • Y9421 M9173 • PF794 M9116 • PF688 V1561 • M9009 Z17352 • Y10881 Y10858 M9064 • PF350 M9377 • PF1241 M9398 • PF1279 M9410 M9097 • PF672 M9397 M9271 M9341 • PF1072 L969 Y9449 V1530 • M9008 Y10846 M9252 V235 • PF1410 M9316 • PF1034 Y10867 M9354 M9049 M9010 M9340 M9129 • PF707

CT M5745 • CTS8608 M5708 A5213 • Y1819_2 M5687 • CTS5019 L1492 • Y1569 M5589 • PF212 PF1415 • V226 • M5603 Y1510 M5657 M5688 Y1544 • Y1544_1 M5608 • PF258 M5751 • PF937 M5821 • PF1269 M5651 • CTS1996 Y1441M5588 • PF210 M5743 • CTS8542 M168 • PF1416 V2824 • M5671 • CTS3662 • PF704 M5593 M5810 • CTS11408 Y1508 M5699 • PF803 PF1016 M5753 • CTS9458 • PF947 Y1528 Y1581 Y1511 M5752 • CTS9296 M5742 • PF904 V1863 • M9022Y1446 M5781 • PF1040 PF110 Y1575 • Y1575_1 Y1475 M5661 • CTS2842 M5786 • PF1061 Y1490 M5769 • PF996 Y1498 M5724 • PF866 M5785 M5777 Y1828 M5775 Y1482 M5609 M5718 • CTS7257 M5707 • CTS6383 M5726 Y1556 • Y1544_2 M5780 L1028 • CTS4368 • M5680 M5716 • PF840 M5725 • CTS7741 • PF867 M5759 Y1485 Y1507 M5738 PF1418 • V52 • M5721 M5798 M5732 • CTS8089 Y1503 M5660 Y1531 V3648 • M5760 • PF954 M5811 • PF1238 Y1571M5613 M5692 M5683 • CTS4650 M5768 M5805 • PF1227 Y1489 V1494 • M5615 • PF269 Y1509 Y1817 M5584 • CTS543 • PF206 M5715 • CTS6907 • PF833 PF192 Y1505 M5670 • CTS3460 Y1452 L1462 • M5587 Y1579 M5597 • CTS1217M5819 Y1448 M5803 M5747 Y1471 M5600 Y1455 V1052 • M5576 • CTS125 Y1496 Y1460 V3623 • M5754 M5686 Y1546 • Y1546_1 Y1585 M5647 Y1465 M5717 • PF844 M5791 • PF1080 M5825 Y1464 M5617 • PF274 Y1574 M5614 • PF266 M5630 M5629 V3641 • M5757 • CTS9555 Y1494 Y1567 M5648 Y1578 Y1594 M5817 M5606 • PF256 M5830 • CTS12633 •

PF1329 Y1450 M5831 Y1488 M5795 M5712 M5813 Y1504 M5772 • CTS10512 V3310 •
M5748 • CTS8980 • PF928 V3642 • M5758 • CTS9556 M5607 V4106 • CTS10362 • PF998 •
M5770 M5801 M5808 • CTS11358 M5636 M5665 • CTS3216 Y1457 Y1524 M5784 •
PF1059 Y1474 M5616 • PF272 M5621 M5656 V1431 • M5612 Y1492 M5720 •
CTS7482 Y1480 Y1521 M5746 • CTS8709 M5578 M5767 • CTS10110 Y1527 M5641 V3337
• M5750 • CTS9014 M5796 • PF1097 Y1470 CTS5457 M5723 • PF862 M5832 •
PF1333 M5741 M5653 • CTS2077 • PF657 M5590 • PF216 CTS1254 • M5598 M5697 •
CTS5746 M5800 • PF1203 Y1589 PF1413 • V189 • M5577 M5762 • CTS9722 Y1568 M5705
• CTS6327 • PF811 PF500 Y1506 Y1593 V1043 • CTS109 • M8948 M5809 •
PF1237 Y1552_2 M5713 • CTS6800 Y1502 M5618 Y1538V2216 • M5633 M294 V3858 •
PF970 M5776 • PF1029 M5711 • PF821 M5794 • PF1092 Y1587 M5602 •
PF246 Y1467 M5802 V2901 • M5675 • PF719 M5591 • PF223 V1325 • M5605 L1480 •
V3908 • M5766 M5783 Y1497 Y1599 Y1473 Y1819 • Y1819_1 M5698 • PF796 M5816 •
CTS11827 Y1525 Y1562 • Y1559_2 M5797 • PF1098 M9150 •
PF750 PF342 M5649 Y1454 Y1438 Y1447 V2175 • M5632 M5709 PF1414 • V9 •
M5585 M5662 V3758 • M5764 • CTS9828 •
PF964 Y1476Y1491 M5601 Y1472 M5628 M5765 Y1495 M5611 •
PF263 Y1461 PF1276 M5626 M5689 M5642 M5700 • CTS6252 M5599 • PF234 V4162 •
M5771 Y1456 M5594 • CTS1109 M5722 • CTS7517 Y1443 M5582 • CTS401 •
PF202 M5736 • CTS8243 • PF891 M5730 Y1586 Y1573 Y1518 M5788 M5814 FGC28144 •
V3808 • Y1539 M5739 • PF898 PF1417 • V41 • M5695 Y1591 M5645 M5659 •
PF667 M5691 • PF779 M5729 • CTS7936 M5822 M5664 • CTS3120 • PF683 Y1462V3317 •
M5749 M5778 • PF1031 M5681 V1401 • M5610 PF1420 • V55 V1653 •
M5625 M5650 Y1440 Y1791 M5737 • PF892 Y1580 Y1590 M5706 • PF815 Y1577 M5792 •
PF1088 M5678 • PF725 Y1514 Y1559 • Y1559_1 M5622 Y1469M5826 CTS11575 •
PF1245 M5620 M5583 • CTS423 V3728.1 • M5763 • CTS9760 • PF961 M5640 •
PF318 M5714 • CTS6890 M5790 V1878 • M5631 • PF292 M5719 • PF850 M5652 •
PF652 Y1483 M5627 Y1451 M5639 Y1444 M5684 • CTS4740 • PF751 M5694 •
CTS5532 M5782 • PF1046 Y1458 M5735 • CTS8166 M5690 • CTS5318 Y1449 M5638 •
PF316 M5679 • CTS4364 M5624 P9.1 • P9 M5728 • CTS7933 M5676 •
PF720 Y1526 M5682 M5818 • CTS11991 M5812 PF15V1540 • M5619 • PF278 M5669 •
CTS3431 M5595 • CTS1181 M5727 • CTS7922 • PF875 M5774 M5804 •
CTS10946 Y1537 M5646 M5756 • PF951
CF M3711 • CTS6376 • PF2697 CTS3818 • PF2668 • M3690 V3489 • PF2723 • M3727 •
F2841 P143 • PF2587
F PF2651 • F1704 • M3675 Y1812 L498 • PF2707 • M3717 CTS4969 • PF2682 •
M3700 PF2743 • M3747 FGC2057 • Y1810 PF2615 • M3652 PF2756 CTS11819 • PF2766 •
M3761 M213 • P137 • PF2673 • PAGES00038 P316 • PF2696M3693 • CTS4139 •

PF2672 M3771 • CTS12632 • PF2775 YSC0001298 • PF2620 • F1302 • M3656 FGC2055 • Y1805 PF2744 • M3748 PF2653 • F1714 • M3677 P138 • PF2655 PF2639 M3696 • CTS4443 • PF2677 PF2611 • M3647 PF2608PF2660 • F1767 • M3683 L468 • PF2689 • M3703 PF2588 • M3635 • CTS71 M3692 • CTS3996 • PF2671 CTS2041 • PF2652 • M3676 M3640 • CTS540 • PF1506 PF2740 • M3744 L929 • PF2605 • M3643 M3687 • CTS3195 • PF2664 CTS3536 • PF2666 • M3688 V1644 • L313 • PF1426 • M3651 PF2739 • M3743 P151 • PF2625 M3728 • CTS9280 • PF2724 M3756 • CTS10983 • PF2760 CTS7981 • PF2710 PF2601 • M3641 • CTS608 CTS5432 • PF2687 • M3702 M3720 • CTS8014 • PF2711 PF2592 PF2637 • M3672 P158 • PF2706 CTS1468 • PF2607 • M3644 PF2742 • F3254 • M3746 L882 • PF2745 • M3749 CTS10290 • PF2735 • M3739 CTS11726 • PF2765 M3730 • CTS9372 • PF2725 CTS12138 • PF2774 • M3770P166 • PF2702 PF2718 • F2710 • M3723 PF2729 • F2964 V3940 • PF2732 • F2993 • M3737 L352 • PF2728 • M3734 M3706 • CTS5948 • PF1695 F3561 • M3766 Y1808 PF2630 • M3664 PF2593 P161 • PF2719 CTS3868 • PF2669 PF2700 • F2402 • M3714 PF2737 • F3136 • M3741 P163 • PF2686 CTS12027 • PF2768 • M3763 M3689 • CTS3654 • PF2667 M3729 • CTS9317 • PF1767 P160 • PF2618 FGC2061 • Y1799 FGC6229 • Y1807 L1074 • CTS4267 • PF2674 • M3694 P14 • PF2704 L851 • CTS11821 • PF2767 • M3762 PF2647 PF2612 • M3648 P140 • PF2703 F3556 • M3765 • PF1914 M3718 • CTS7878 FGC2046 • Y1809 CTS4557 • PF2679 • M3698 PF2629 • M3663 V3268 • PF2683 • F2048 • M3701 PF2747 • M3750 PF2688 • F2142 PF2609 • M3645 P159 • PF2717 V2513 • L543 • PF2663 • M3686 P135 • PF2741 CTS11150 • PF2761 • M3758 P157 • PF2771 CTS1932 • PF2650 • M3674 CTS6135 • PF2693 • M3708 PF2736 • F3111 • M3740PAGES00080 • M235 • PF2665 • PAGE80 P187 • PF2632 PF2627 • M3661 P139 • PF2698 M3732 • CTS9456 • PF1438 M3731 • CTS9418 • PF2726 PF2594 M3712 • CTS6542 • PF2699 V1990 • YSC0001297 • F1209 • M3654 L350 • PF2692 • M3707 PF2750 • M3753 CTS9534 • PF2727 • M3733 M3725 • CTS8985 • PF2721 FGC2054 • Y1811 V3919 • PF2731 • F2985 • M3736 FGC2049 • Y1802 FGC2069 • Y1758 F3512 • PF1911 M3699 • CTS4737 • PF2680 P148.2 • P148.1 • P148 • PF2734 FGC2062 • Y1820 PF2628 • M3662 PF2590 • V205 • M3638 F3584 • M3768 • PF1916 Y2888 F773 P149 • PF2720 V1597 • PF2614 • F1089 • M3649 CTS11370 • PF2763 • M3759 M89 • PF2746 PF2591 • M3639 V3900 • L470 • PF2730 • M3735 M3721 • CTS8467 • PF2715 P133 • PF2636 FGC2092 • Y4457 PF2589 • V186 • M3637 IMS-JST003305 • V1029 • F719 • M3636 Y1800 PF2600 PF2631 • M3665 CTS2097 • PF2654 • M3678 F3692 • M3650 PF2713 PF2598P136 • PF2762 L132.1 • L132 • L132.2 • PF1437 M3713 • CTS6843 • PF1720 V1355 • YSC0001295 • PF2610 • F1046 • M3646 M3716 • CTS7301 • PF2705 P142 • PF2604 CTS7002 • PF2701 • M3715 PF2613 CTS5264 • PF2684 V2194 • PF2619 • F1285 • M3655 P141 • PF2602 PF2621 • F1320 • M3657 PF2758 • F3335 • M3754 M3724 • CTS8638 P146 •

By James Lee Rader

PF2623 PF2772 • F3616 • M3769 P134 • PF2606 P145 • PF2617 M3666 • PF1580 PF2749 • M3752 PF2616 • F1149 • M3653 Y1813 M3682 • CTS2480 • PF2659

GHIJK M3684 • CTS2569 • PF2661 M3680 • CTS2254 • PF2657 V2308 • YSC0001299 • PF2622 • F1329 • M3658 M3773 • CTS12673

HIJK PF3494 • F929 • M578

IJK L16 • M522 • S138 • PF3493 V1438 • YSC0001319 • PF3497 • M2684 L15 • M523 • S137 • PF3492 • Z4413 PF3500 • M2696 V1295 • PF3495 • F3689 • M2682 • Y2571

K P132 • PF5480 PF5500 • F2548 • M2692 L819 • CTS4265 • PF5494 • M2686 V3169 • PF5495 • F2006 • M2688 PF5470 A5331 • YSC0000055 • PF5459 • M2348 M2352 • CTS2071 • PF5489 P131 • PF5493 L469 • PF5499 • M2689 PF5469 • V104 M9 • PF5506 PF5488 • M2351 V4038 • PF5503 • F3026 • M2694 P128 • PF5504 YSC0000222 • PF5505 • L1346 • M2695 CTS9278 • PF5501 • M2693 CTS10976 • PF5509 • M2698

K(xLT) M526 • PF5979

MP-M1205 PF5852 PF5969 L405 • PF5990 M1205 • CTS2019 P331 • M1221 • YSC0000186 • PF5911

P P237 • PF5873 CTS12028 • PF5977 • M1272 F313 • M1219 YSC0001257 • CTS1907 • PF5894 • M1204 M1235 • CTS7604 • PF5928 M45 • PF5962 PF5849 Y503 • Y503_1 L741 • PF5960 • YSC0000297 M1149 • CTS10168 • PF6061 M1270 • CTS11173 • PF5974 P235 • PF5946 PF5878 • M1194 PF5985 PF5945 • F524 • M1248 PF5991 • F4 • M1183 M1228 • CTS5884 • PF5917 V1809.1 • PF5872 • M1192 CTS3316 • M1209CTS5808 • PF5915 • M1226 Y451 PF5854 PF5914 • F332 • M1224 Y272 M1264 • YSC0000227 PF5483 PF5871 • M1190 PF5886 • M1199 92R7_1 P228 • PF5927 Y44 • M1207 PF5975 • F640 • M1271 PF5942 • M1244 • CTS9604 CTS6948 • PF5922 • M1231 V1195 • PF5861 • F83 • M1185 V2974 • L779 • PF5907 • YSC0000251 YSC0001285 • CTS5673 • PF5497 • M1225 P230 • PF5925 CTS3775 • PF5906 • M1214 L768 • PF5976 • YSC0000274 PF5848 P283 • PF5966 PF5964 • M1263 PF5958 • M1160 P226 • PF5879 P295 • S8 • PF5866 CTS3736 • PF5905 • M1213 PF5892 • M1202 PF5876 • M1193 CTS1518 PF5951 • F556PF5901 • F1857 • PAGES00083 • PAGE83 FGC216 • Y448 PF5461 PF5993 • Z1244 V231 • PF5862 • F91 PF5891 PF5881 • F180 • M1196 Y444 PF5956 • M1259 Y45 • M1208 CTS3697 • PF5904 • M1212 M1186 • YSC0000279 • PF5864 V1651 • PF5870 • F115 • M1189 V2979 • M1216 • YSC0000176 • PF5908 P243 • PF5874 Y504 • Y503_2 L82 • PF5972 PF5994 V3529 • F506 • PF5940 • M1243 • YSC0000966 PF5916 • F344 • M1227 M1240 • YSC0000205 • PF5936 L721 • PF6020 PF5887 V3732 • M1246 • YSC0000270 • PF5943 PF5882 CTS5418 • PF5912 • M1222 FGC213 • Y455 PF5846 • M1184 • CTS216 V1079 • PF5845 • CTS196PF5954 • M1256 PF5970 Y269 • Y269_1 PF5984 • F680 • M1275 PF5944 • F521 • M1247 P282 • PF5932 PF5957 • M1260 PF5867 CTS7886 • PF5929 • M1236 Y450 V3240 • M1218 • CTS4944 • PF5909 P284L781 • PF5875 • YSC0000255 PF5850 PF5888 P244 • PF5896 •

P244.1 • P244.2 P240 • PF5897 L1185 • CTS9162 • PF5937 • M1241 PF5869 • M1188 PF5971 PF5978 • F647 • M1273 FGC286 • Y456 PF5900 • CTS3358 • M1210 • PF5899 CTS12299 • PF5987 P69 FGC211 • Y447 PF5885 • F212 • M1198 PF5883 • M1197 L268 • PF5983 P27_1 • P27.1_1 • P207 • P27.2_1 PF5853 L138 PF5855 PF5880 • M1195 Y446V4004 • PF5949 • F536 • M1251 FGC83 • Y1403 P281 • PF5941 PF5980 • F653 Y483 • Y483_1 M1269 • CTS10859 FGC217 • Y458 M74 • N12 • PF5963 PF5851 L536 • PF5860 PF5955 • M1257 PF5981 • F671 • M1274 P239 • PF5930 L471 • PF5989 PF5982 L821 • PF5857 • F29 • M5579 PF5471 PF5473 M1250 • CTS10085 • PF5948 M1109 • CTS4437 CTS3813 • PF5491 • M1215 PF5865 • M1187

R M732 • CTS8311 • PF6055 CTS3622 • PF6037 YSC0000201 • PF6057 • M734 • S4 P280 • PF6068 P227 FGC202 • Y453 M696 • CTS5815 • PF6044 P232 P224 • PF6050 YSC0000232 • M789 • L1225 • PF6076 FGC201 • P285 • PF6059 CTS2913 • PF6034 • M667 V2573 • YSC0001265 • CTS3229 • PF6036 • M672 PF5953 • M764 M207 • UTY2 • PF6038 • PAGES00037 V3466 • CTS9200 • PF5938 FGC208 • Y457Y480 YSC0000233 • PF6077 • L1347 • M792 P229 • PF6019 Y472 • F47 • M607 • PF6014 • S9 M795 • CTS11075 • PF6078

R-Y482 FGC1168 • PF6040 • YSC179 M799 • PF6079 • YSC237 PF5919 • F356 • M703 Y482 • PF6056 • F459

R1 M813 • CTS12618 • PF6089 L875 • PF6131 • YSC0000288 • M706 M730 • CTS8116 • PF6138 P231 P236 • PF6137 M717 • CTS7122 • PF6135 P245 • PF6117 P233 • PF6142 FGC465 • PF6146M173 • P241 • PF6126 • PAGES00029 FGC218 • Y464 • PF6008 PF6069 M781 • PF6145 CTS3321 • PF6125 • M673 P286 • PF6136 PF6073 CTS2680 Y512 FGC189 • Y305 • PF6031 Y290 • F211PF6118 • M640 M812 • CTS12546 • PF6088 PF6110 FGC193 • PF6011 M663 • CTS2565 • PF6122 P234 • PF6141 V1478 • PF6116 • F102 • M625 PF6133 • F378 • M711 FGC198 • Y465 YSC0000230 • L1352 • M785 V1356 • PF6114 • F93 • M621 P242 • PF6113 FGC13894 • P294 • PF6112 F132 • M632 M643 PF6119 M748 • YSC0000207 PF5477 • F28 CTS3123 • PF6124 • M670 M306 • S1 • PF6147Y459 P225 • PF6128 P238 • PF6115 M714 • CTS7066 • PF6049

R1b M343 • PF6242
R1b1 L278 M415 • PF6251
R-L389 L389 • PF6531
R-P297 YSC0000269 • PF6475 • S17 FGC69 • L320 • PF6092 P297 • PF6398
R-M269 V1741 • L483 • PF6097 M269 • PF6517 CTS12478 • PF6529 S351 • L150 • PF6274 • L150.1 • PF6274.1 • L150.2 • PF6274.2 L1063 • CTS8728 • PF6480 • S13 S10 • PF6399CTS11468 • PF6520 CTS8665 • PF6479 PF6494 PF6500 PF6438 V3866 • L1353 • YSC0000294 • PF6489 YSC0000203 • PF6482 PF6409 PF6495 PF6404 PF6509 CTS10834L773 • PF6421 •

By James Lee Rader

YSC0000276 YSC0000219 • PF6497 PF6443 PF6430 PF6425 PF6455 •

F1794 PF6507 M12188 • L49 • L49.1 • L49.2 • PF6276 • S349 L482 • PF6427 PF6410 •

M520V1699 • PF6432 CTS2664 • PF6454 CTS3575 • PF6457 S3 • PF6485 L500 •

PF6481 PF6419 • CTS623 L265 • PF6431 L407 • PF6252 PF6434 CTS8591 • PF6477

R-L23 L23 • S141 • PF6534 L478 • PF6403

R-L51 FGC39 • CTS10373 • PF6537 PF6535 L51 • M412 • S167 • PF6536 PF6414

R-L151 PF6416 L11 • S127 • PF6539 PF6538 L151 • PF6542 P311 • S128 • PF6545 L52 •

PF6541 P310 • S129 • PF6546 YSC0000191 • PF6543 • S1159 CTS7650 • PF6544 •

S1164 Z8159 • FGC796 • Y101 PF5856 YSC0001249 • CTS10353 • S1175

R-P312 P312 • S116 • PF6547 Z1904 • CTS12684 • PF6548

R-DF27 DF27 • S250

R-Z195 Z196 • S355 Z195 • S227

R-Z198 Z198 • S228

R-Z292 ZS312 • Y964 • M7953 • Y964.2 • M7953.2 • Y964.1 • M7953.1 Z292 • S460

R-Z262 Z200 • S361 Z691 Z263 Z267 V3130 • Z689 • CTS4716 Z262 M167 •

SRY2627 Z688 FGC11216 • Z265 • PF1185 Z201 • S457 Z204Z199 • S234 Z269

R-Z202 Z264 Z266 Z205 Z202 Z203

R-Z206 Z206 • S235 Z690

R-Z208 Z208 • S362 Z693 Z207 • S251 CTS8289

R-Y15783 Y15784 Y15783 Y15646

Legend: Positive Negative Ambiguouse No call

Made in the USA
Columbia, SC
07 July 2017